REORIENTING
HEALTH SERVICES
Application of a Systems Approach

NATO CONFERENCE SERIES

I   Ecology
II   Systems Science
III   Human Factors
IV   Marine Sciences
V   Air–Sea Interactions
VI   Materials Science

II SYSTEMS SCIENCE

Recent volumes in this series

Volume 5   Applied General Systems Research: Recent Developments
and Trends
Edited by George J. Klir

Volume 6   Evaluating New Telecommunications Services
Edited by Martin C. J. Elton, William A. Lucas, and
David W. Conrath

Volume 7   Manpower Planning and Organization Design
Edited by Donald T. Bryant and Richard J. Niehaus

Volume 8   Search Theory and Applications
Edited by K. Brian Haley and Lawrence D. Stone

Volume 9   Energy Policy Planning
Edited by B. A. Bayraktar, E. A. Cherniavsky, M. A. Laughton, and
L. E. Ruff

Volume 10   Applied Operations Research in Fishing
Edited by K. Brian Haley

Volume 11   Work, Organizations, and Technological Change
Edited by Gerhard Mensch and Richard J. Niehaus

Volume 12   Systems Analysis in Urban Policy-Making and Planning
Edited by Michael Batty and Bruce Hutchinson

Volume 13   Homotopy Methods and Global Convergence
Edited by B. Curtis Eaves, Floyd J. Gould,
Heinz-Otto Peitgen, and Michael J. Todd

Volume 14   Efficiency of Manufacturing Systems
Edited by B. Wilson, C. C. Berg, and D. French

Volume 15   Reorienting Health Services: Application of a Systems Approach
Edited by Charles O. Pannenborg, Albert van der Werff,
Gary B. Hirsch, and Keith Barnard

Volume 16   Adaptive Control of Ill-Defined Systems
Edited by Oliver G. Selfridge, Edwina L. Rissland, and
Michael A. Arbib

REORIENTING HEALTH SERVICES

Application of a Systems Approach

Edited by

Charles O. Pannenborg

Albert van der Werff

Department of Public Health and
Environmental Hygiene
Leidschendam, The Netherlands

Gary B. Hirsch

Health Management Consultants
Wayland, Massachusetts, USA

and

Keith Barnard

Nuffield Centre for Health Services Studies
The University of Leeds
Leeds, United Kingdom

Published in cooperation with NATO Scientific Affairs Division

PLENUM PRESS · NEW YORK AND LONDON

Library of Congress Cataloging in Publication Data

NATO Advanced Research Institute of Health Services Systems (1982: Hague, Netherlands)
 Reorienting health services.

 (NATO conference series. II, Systems science; v. 15)
 "Proceedings of a NATO Advanced Research Institute on Health Services Systems, held August 29–September 3, 1982, at The Hague, The Netherlands"—T.p. verso.
 "Published in cooperation with NATO Scientific Affairs Division."
 1. Medical care—Congresses. 2. System analysis—Congresses. I. Pannenborg, Charles O. II. North Atlantic Treaty Organization. Scientific Affairs Division. III. Title. IV. Series. [DNLM: 1. Health services—Organization and administration—Congresses. 2. Systems analysis—Congresses. W 84.1 N279r 1982]
 RA394.N37 1982   362.1   83-17750

 ISBN-13: 978-1-4612-9670-6   e-ISBN-13: 978-1-4613-2685-4
 DOI: 10.1007/978-1-4613-2685-4

Proceedings of a NATO Advanced Research Institute on Health Services Systems, held August 29–September 3, 1982, at The Hague, The Netherlands

© 1984 Plenum Press, New York
Softcover reprint of the hardcover 1st edition 1984
A Division of Plenum Publishing Corporation
233 Spring Street, New York, N.Y. 10013
All rights reserved

ACKNOWLEDGEMENTS

The realization of a working conference on health
services systems, covering such a large field, depends
on the support of a large number of people.

First of all this NATO-Advanced Research Institute
on Health Services Systems owes success to the dedicated
members of the Programme Committee, who have given their
unsparing support and have so generously contributed their
time. They are:

Mr. G.B. Hirsch, USA
Prof. P.B. Checkland, UK
Dr. D.M. Pendreigh, UK
Prof. M.I. Roemer, USA
Dr. G. Wennström, Sweden
Dr. B.M. Kleczkowski, Switzerland, WHO, advisor

Special recognition is given to our Proceedings Editors:

Dr. Ch. O. Pannenborg, The Netherlands
Mr. K. Barnard, UK
Mr. G. Hirsch, USA

Neither the Conference, nor the Proceedings would have been
possible without the skill, efforts and support given by the
Organizing Committee:

Dr. Ch. O. Pannenborg
Mr. A.H. Zwennes
Mr. P. Käuderer

Discussions during the Conference could not have been rendered
into text by the conference end without the tireless efforts
of the Rapporteurs; they are

of the Ministry of Public Health at the Hague:
Mrs. H. Emanuel

Mrs. L. Gunning
Mrs. Dr. J. Hageman
Dr. J.W. Hartgerink
and of the University of Limburg:
Mr. A.W.M. Meijer
Mrs. Dr. I. Mur

I would also like to specifically acknowledge my staff
both at the Ministry of Public Health and Environmental
Hygiene and at the State University of Limburg who have
assiduously dispatched every assignment to make this
Conference succeed.

A special note of thanks is due to

Mrs. M.A. Holtmans
Mrs. M.A. Lamerée
Mrs. F.W.M.A. Jans de Haan

for their organizational and secretarial support during
the preparation during the Conference itself and especially
during the period after the Conference with regard to the
typing and editing of the underlying Proceedings.

Albert van der Werff
Programme and Conference Chairman
NATO Advanced Research Institute
on Health Services Systems
The Hague, 1982

PREFACE

The Advanced Research Institute on "Health Services Systems"
was held under the auspices of the NATO Special Programme Panel on
Systems Science as a part of the NATO Science Committee's continuous
effort to promote the advancement of science through international
cooperation. A special word is said in this respect supra by Professor Checkland, Chairman of the Systems Science Panel.

The Advanced Research Institute (ARI) was organized for the
purpose of bringing together senior scientists to seek a consensus
on the assessment of the present state of knowledge on the specific
topic of "health services systems" and to present views and recommendations for future health services research directions, which
should be of value to both the scientific community and the people
in charge of reorienting health services.

The conference was structured so as to permit the assembly
of a variety of complementary viewpoints through intensive group
discussions to be the basis of this final report. Invitees were
selected from all over Europe and North America, including observers
from the major international organizations involved in health services research, to provide the experience and expertise necessary
to make the issue of "health services systems" valid and significant.

The structure of the ARI was designed along the six topics
mentioned and highlighted in the introductory address "Health Services Systems in the Western World" (p. 7). The meeting was basically organized in two modes: plenary sessions and parallel workshops on specialized topics. Intensive personal contacts and evening sessions where reports from the day's workshops were presented
proved to be very useful for the better understanding of the problems and needs of health services with regard to the systems approach. Rapporteurs who were assigned to the plenary systems and
the chairman and rapporteurs of each workshop prepared summaries
which provided the basis for the reports that were ultimately drafted.

Special thanks are due for the help and assistance of Dr. B.A.
Bayraktar of the Scientific Affairs Division, NATO, which so generous-

ly provided the ways and means to organize and execute the Advanced
Research Institute in such a broad manner. The same holds true for
the Ministry of Public Health at the Hague, specifically its Director-General Dr. J. van Donden, who provided the conference equally
with both financial and support in natura, for which a special ack-
nowledgement is in order.

The ultimate success of any conference is based on the activity
of the participants and observers, and they in particular deserve
our special thanks, not only for the excellent presentations and ultimate manuscripts which they prepared, but for their devotion to the
task of interchanging ideas on a topic which has important implications for the future restructuring of many national and regional
health services and ultimately profound consequences for the health
and health care of individuals and communities.

Because of the intensity and enthusiasm of the participants
and observers, who make the ARI a most memorable one, the real success of the conference cannot be determined only by the quality of
this volume, but by its impact on future work in this new and important area of health services research.

April 1983   Charles O. Pannenborg
The Hague   Editor

CONTENTS

Special addresses

Opening address of the
NATO ARI-HSS Conference   1
   M.H.M.F. Gardeniers-Berendsen

Adress from the Systems
Science Panel...............................   5
   P.B. Checkland

Health Services Systems in the
Western World: An International
Perspective on Problems and Prospects..........   7
   A. van der Werff

SECTION I   Re-orienting Health Services-Application of
         a Systems Approach

A Systems view of Health Services and Their
Reorientation: Summary of the Conference
Proceedings and Background Papers...............   15
   G.B. Hirsch

Analysis of Health Services Systems-
A General Approach   47
   M.I.Roemer

"A Systems Approach" and "Health Service
Systems": Time to Re-think?.....................   61
   P.B. Checkland

SECTION II   Tools for doing it

Introduction...................................   67
   A. van der Werff

Some Aspects of Alternative
Approaches to Planning............................   69
   A.W.M. Meijer

Manpower Planning and Health
Services Systems..................................   91
   G. Wennström

Financing and Health
Services Systems..................................   99
   L. Delesie

Cost-effectiveness and Health
Services Systems -
A Systems Approach to the
Assessment of Health Care.........................   115
   C.O. Pannenborg

Health Services Systems and
Information Services..............................   137
   A.S. Härö

Research and Development
and the Health Systems Approach...................   153
   D. Affeld

SECTION III   National Experiences

Introduction.....................................   165
   A. van der Werff

The French Health Services
System...   167
   J.F. Lacronique

The Dutch Health Services System:
Problems and Measures of Interdepen-
dencies and Control..............................   179
   A.C.J. de Leeuw
   I.M. Mur-Veeman

The Italian Health
Services System..................................   203
   A. Bariletti

Health Services in the United States:
Groping Toward a "System"?.......................   219
   A.R. Somers

CONTENTS

The Health Services
Systems in Belgium.............................   249
  J.E. Blanpain

The Health Services
System in Turkey...............................   255
  M.R. Dirican

The Health Services of Norway:
on the Road to a Political System..............   271
  H.T. Skaug

The Canadian Health Services System............   283
  A. Crichton

The Health Services System of the
United Kingdom.................................   309
  K. Barnard
  D.M. Pendreigh

The Changing Health Services System in Greece....   347
  A. Ritsataki

Participants...................................   357

Index..   365

OPENING ADDRESS OF THE NATO ARI-HSS CONFERENCE

M.H.M.F. Gardeniers-Berendsen

Minister of Health and Environmental Protection

It is a great pleasure to me to open this Conference on Health Services Systems, that all of you will attend this week. Most of you came from far away and I would like to welcome you all to our country The Netherlands and to the residence of our Government here at The Hague.

Whilst the Conference is sponsored both by the NATO Panel for Science Systems and by the Ministry of Health and Environmental Protection, we are very happy to act as your host for this important Meeting.

The subject that you will discuss is important to us. The health services in most Western countries have evolved up to the point where, as one of your colleagues once said, we could almost speak of a "non-system". The expenditures for the health services in some of our countries account for almost 10 percent or more of the Gross National Product. Increasingly, serious and disconcerting questions are being asked about the effectiveness of our health services. To what extent do they contribute to better health and to less disease of our populations? Are other factors, such as environmental hygiene, urbanization or cultural customs more important in determining the health of our nations? Have our health services grown to their present size just only because of the open-ended social security systems and not because of their intrinsic marginal value? Does the demand by the health services for monies from the capital-market inhibit or damage our capacity to generate the productive investments that are so necessary in these times of economic decline?

From such questions, it is clear that none of our Governments these days can afford to treat its health services haphazardly. In order to manage our health services properly and to plan them according to their social and economic value a systematic approach has to be taken. It will not do anymore to organize the health sector on an ad-hoc basis, mainly responding to problems if and when they arise, without actively pursuing a systematic policy.

Ladies and Gentlemen, you may wonder why I tell you all these things, now, August 1982, and whether I should not have told you so already several years ago. Or, in short, you may silently address your neighbour and whisper "now, I could have told you so several years ago!"
Well, I can tell you that here in Holland, indeed we said so several years ago.

Starting in 1974, we made our first serious attempt to create the necessary structure that would enable us to plan and to manage our health services in a more systematic way. The White Paper on the Structure of the Health Services identified the various sub-systems operating in our health sector; it proposed ways and means to come to grips with the often autonomously functioning elements of our health sector.

As you will appreciate, to restructure the health sector into a logical and comprehensive system which can be controlled within the limits of available resources requires a long-term effort. We made the effort and are still very much in the process of introducing the systems approach into our health sector. A Government cannot do so without the necessary legislation and political support to make such an endeavour realistic.

Under the new Hospital Facilities Act, we started a more systematic process of our hospital planning. The recently approved Law on Health Care Tariffs is meant to produce a more manageable and logical structure of our health care finances. And at this very moment we have the Final Reading in Parliament of the new Health Services Act, which will enable us - we hope - to plan our primary care, our secondary care and our tertiary care in such a way that their interrelationships constitute a meaningful and sensible sequence.
Finally, we propose to introduce a Bill on the Health Professions, which will create a more sytematic structure with regard to the numerous medical and health professions of today.

Some of the key concepts of this conglomerate of Laws to bring about a comprehensive and consistent system of health services, are:

- Cohesion, by which we mean to establish a much better and more meaningful coordination between and among the various health services, social services and community activities that now often operate independent of each other. The present absence of integration between the geriatric care provided by the family practitioner, then by the geriatric ward of the hospital, and ultimately by the psycho-geriatric community-home is a vivid and often embarrassing example.

- Another concept is regionalization. Although you may wonder how it is possible to regionalize such a small country as Holland, we have learnt valuable lessons from some of the countries that you represent here today. To involve local and regional health authorities in their own health planning, to decentralize the principal decision-process to the level where the actual health services occur, these are concepts by which the translation of our policies into effective operational health plans can be much better guaranteed than at present is the case.
- A third concept we find important is democratisation of services. In order to ensure a just position of consumer and patients' interests within the process of health policy determination and health services delivery, new systems of consumer participation have to be designed. The issue is rather complex, entailing such questions as "informed consent", financial solidarity, or the specification of the "right to health care"; examples from some of your countries like the "community health centre" approach or the "health maintenance organisations" may well be helpful as guidelines for our system.

How, then, can a Working Conference like this contribute to the clarification and better implementation of our objectives?

It is my opinion that all the Laws that we enact and all the concepts that we endorse run the risk of remaining misunderstood, if at the same time we don't clearly identify and thoroughly clarify all the various sub-systems that determine the reality of day-to-day health care. In this respect I like to mention to you some of the health care systems I consider to be of crucial importance.

Already I referred to the need for a comprehensive structure, which encompasses all the relevant health care functions and spells out their mutual relationships in a consistent manner.

Then, we need a clear picture of the resources for the health services, among which we regard the pressing problem of health manpower: on the one hand the powerful system of health manpower education with the Faculties of Medicine, the professional societies, and the right to an education of one's own choice; on the other hand the overproduction of certain health professionals, increasing unemployment of physicians and the absence of an effective relationship with the need for certain health professions.

A third area of interest is the sub-system of economic and financial support of the health services. Often we see that this is functioning according to its own rules and procedures with regard to insurance, social security premiums, investments and budgeting-and-accounting. Its relationship with clinical decisionmaking, average length of hospital-stay or utilization-rates of drugs remain weak and obscure.

Other sub-systems of the health sector that would need your attention are those of health care <u>delivery patterns</u> and of health services <u>management</u>.

Ladies and Gentlemen, I hope that I have made it sufficient-ly clear to you that what we need in most of the Western health sectors, is a better understanding and a greater awareness of what kind of seperate health services systems are at work. To clarify whether they work at cross-purposes and to propose solutions which will link them together. Only if the various components of our overall health system are consistently geared to each other, it will be possible to redirect our health ser-vices to become more effective and more efficient.

I wish you a successful Conference!

AUGUST 30,
1982

ADDRESS FROM THE SYSTEMS SCIENCE PANEL

P.B. Checkland

Chairman of the NATO-Systems Science Panel

NATO is best known as a military alliance, but from its earliest days has funded a civilian Science Programme. The argument of the founders was that the Alliance would be strong if the science of its member countries were strong. Given the nature of science, the programme is of course "open"; contributers to meetings speak as individuals, and proceedings are always published in the general literature.

Within the Science Programme the Systems Science Panel has for more than a decade organised meetings and exchange-visits, funded research, and paid student grants to enable students from one member country to study in another. Its area of activity has always been broadly conceived, and its activities have included not only support for technical work in systems science but also support for meetings at which experts reflect upon and discuss important real-world problems areas - such as transportation, health care and industrial development - to which systems thinking is relevant.

In the final year of its programme, the Systems Science Panel has been very pleased to provide funds to support a meeting to discuss the problems of Health Service Systems, a problem area of considerable importance at a time of increasingly aged populations and reduced government expenditures on social services.

HEALTH SERVICES SYSTEMS IN THE WESTERN WORLD:

An International Perspective on Problems and Prospects

Albert van der Werff

Ministry of Health and Environmental
Protection, Leidschendam

National University of Limburg
Maastricht, The Netherlands

Background of the study

Major Concerns With Respect To The Development Of The System-Concept Of Health Services In Western Countries

In most Western countries health care has attained an acceptable level. The state of medical science and technology renders it possible to supply extensive services. In most of these countries the health services has developed into one of the largest sectors of the economy requiring up to 5-10% of the national income and utilizing approximately 4-6% of the total working population. In those countries too one finds the same accelerated cost increase, which exceeds the rise in national income and which is largely caused by the everincreasing demand for therapeutic services in hospital care. In addition the health sector shows weakness from the systems' point of view. As a consequence the health sector increasingly gets public attention.

A weak development of the system concept has been, and still is, a major concern in many countries. Without trying to be complete, the following shortcomings can be listed in general:

Health Services Systems' Environment

National Health Services Systems may not be fully in line

with their environments within which they have to carry out their
functions. As a consequence, the goals and objectives of National
Health Services Systems may inadequately meet the values and requirements of society.

Health Service System Structure

The health services may lack an adequately structured and
professionally guided health administrative and management system
at one or at all levels. The co-ordination within the health sector
and, in particular, between the public and private sector may be
weak. Also the co-ordination of the health sector with other (related)
sectors may be insufficient.

Health Resources Production

The health resource production with respect to manpower, including education, facilities and knowledge may inadequately be
used or insufficiently be related to the needs of the population
as well as to the patterns of delivery, organization and functioning
of the health services systems.

Economic, incl. Financial Support Of Health Services Systems

Health expenditures may grow disproportionally in comparison
with other sectors and exceed the economic potential of the country.
Instruments for the control of cost development, both with respect to
the whole of the health services system and to its components, are
still inadequate or even partly lacking. The structure of the financing of health care may be too fragmented and does not promote the
necessary collaboration between the services. Because the financial
mechanisms are frequently too passively adapted to the existing organizational situation, they fail to exert corrective influences on
the co-ordination of the various service units which are composing
the system. At the same time the burden of health care expenditures
on different population groups may be unequally distributed.

Patterns Of Health Care Delivery

The health service delivery may insufficiently be in tune with
the real needs of the population and may neglect particular population groups. Also preventive care may be neglected, and in contrast, personal curative care be overemphasized. The delivery of the
health services may be out of balance at various points. As a result of the medico-technical development usually a disproportionate
large stress comes to fall on the hospital sector whereas primary
health care is often lacking an adequate infrastructure. The delivery pattern of health services is frequently also displaying insuf-

ficient cohesion, the result being independent functioning of services alongside one another. Often the co-operation between the health
services and social (welfare) services is weak too, or even absent.

Health Service System Management

The leadership and managerial skills to run a Health Services
System in its present form may be well developed. In addition, one
may observe that the managerial capacity for innovation on the basis
of a systems view may be inadequate. Moreover, the health management
processes and mechanisms as health policy formulation, planning, monitoring and evaluation, information processing and health system
research may be weak. Frequently mechanisms of regulation are also
incomplete as well as the related legislation.

Developments Of Health Services Systems In Different Western Countries

In order to cure this situation many countries are in the process of improving the system's concept of the health services, the
aim being to reorganize the health services into a coherent whole,
whose effect on health and related costs can be controlled. Usually
this process involves a decentralization of the administrative system. This activity is of great complexity as due account has to be
taken of a multiplicity of factors which may change in the course
of time.

Furthermore, the group affected by organizational action in
health care is large and heterogeneous. The organization of Health
Services Systems will therefore demand considerable time during
which the circumstances to which the Health Services Systems are
adapted will change. Thus, a situation of full equilibrium can rarely be achieved. For this reason the problem of organizing Health
Service Systems may be regarded as a question of development. It can
be noticed that different countries are in different phases of development.

In some countries (as in Denmark, Norway, Sweden and Finland)
the "decentralized" political/administrative structure has facilitated the establishment of "highly organized systems" in an early stage,
which have shown a considerable improvement during the seventies.
In 1974, Great Britain created a "deconcentrated" administrative
hierarchy which was considered an important requirement for the
unification of the different components of the health services system.
The national health systems of other European countries could be
characterized as "moderately organized". In many of these countries
a process of reorientation of the health services has been initiated in
the beginning of the 1970's. Some countries are intending to strengthen
the health services by organizing them as "systems". The Netherlands,

e.g., had developed legislation which has to serve as the basis for
the restructuring of the health services on the systems concept.
Italy, Portugal and Turkey are following this pattern too. Also the
Greek government now intends to reorganize the health services as a
whole and to bring this sector under government control. France,
whose national health system could also be characterized as "moderately organized", has gradually reduced the pluriform element of
the existing complex. Other countries, however, such as Germany,
and certainly also Belgium, have been reluctant to strenghten the
degree of formal organization of their national health services.
Also in North America, similar developments may be observed. In the
beginning of the 1970's Canada initiated a process of reorientation
of its health services adopting a systems view with respect to
health care; in addition, regional management has been created within the provinces on a district basis. The health services in the
U.S. have traditionally been organized in a highly "individualistic"
fashion. Reorganization of the existing health services in a particular geographical area under the leadership of government, as could
be carried out in most of the European countries, in general is not
feasible in the U.S. However, also in the U.S. developments of the
organization for comprehensive health care have been initiated.
Examples are the grouping of community health facilities around a
community hospital and health maintenance organizations.

Developments With Respect To The System Concept Of Health At The International Level

In particular at the international level the problem of a weak
systems concept of national health services systems has been given
due attention. In 1973, the Executive Board of the World Health Organization (W.H.O.), following a careful study of the world health
situation 1) came to the conclusion that in many countries health
services were seriously deficient from the systems point of view
and the weaknesses which were listed earlier were all recognized.
In particular, W.H.O. has stressed the point of view that the development of National Health Systems is not an affair of the traditionally defined health sector and its "health services" alone
but is closely interlinked with all aspects of national socio-
economic development. It has been recognized that improvements in
the health of populations can only be achieved as a result of a
strong political will, co-ordinated efforts of the health and
other health-related sectors and conscious involvements of communities.

The comprehensive approach to the development of National
Health Systems has been further stimulated by three crucial events:

o   adoption at the 1977 World Health Assembly of the
     ' Health for All by the Year 2000 ' concept as a mutual
     goal of W.H.O. and all Member States;

o   formulation by the 1978 International Conference at Alma-
     Ata of the concept of Primary Health Care as a leading
     strategy in achieving ' Health for All ' 2);

o   adoption at the 1981 World Health Assembly of the ' Global
     Strategy for Health for All by the Year 2000 ' 3).

The first two declarations stated, that there will be an even distribution among the populations of whatever resources for health are
available (nationally and internationally) and that essential health
care will be accessible to all individuals and families in an acceptable and affordable way, and with their full involvement, so that
all citizens of the world will have the opportunity to attain, by
the year 2000, a level of health that will permit them to lead a
socially and economically productive life.

     The latter declaration of the ' global strategy ' explicitely
indicated that achievement of the ' Health for All ' goal will
require relevant ' reorientation of National health systems ',
so as to develop their appropriate organizational infrastructure
based on primary health care.

     At the same time, the W.H.O. recognized that reorientation of
national health systems will require a fairly simple but scientific sound and well-organized knowledge for those responsible for
system design and development at country level. This knowledge
should encompass basic health systems components, their structural
and functional interrelations, the political and economic conditions
influencing their development, and the possible mechanisms for initia-
tion and maintenance of ' reorientation processes ' towards the desired change 4).

     Similar support for improvement of the system concept of
National health services came from the UN, the Council of Europe,
the European Economic Community, the Centre for Educational Research
and Innovation (CERI) of the Organization for Economic Co-operation
and Development, as well as from the International Institute for
Applied System Analysis.

Objectives Of The Study

     The practical activity of organizing health services systems
has been undertaken now for a number of years in many countries.
Taking into account the need for improvement of the available

knowledge of those involved in health system design and development, as expressed a.o. by W.H.O., it may be of great importance at this point of development to look back on the ' seventies ' and to evaluate the progress made and to exchange experiences in order to be able to review the current approaches, methods and techniques or instruments if this would prove to be required.

On the basis of what was said above, the overall objective of this study was defined as an exchange of experiences on the general systems approach as a tool to build and reorient the health services with a view to ensuring its proper functioning in its environment and to improve its effectiveness within the limits of the available resources.

The specific objectives are:

o   to analyse the accumulated experience in developing
     and reorienting Health Services Systems;

o   to list the shortcomings and obstacles and their influence
     on the process of achieving more coherence in the existing
     health services complexes;

o   to review the current methodologies, mechanisms and instru-
     ments in use, with a view to improving them, special atten-
     tion being devoted to health policy formulation, health
     planning, monitoring and evaluation, health system research,
     information processing, regulation and the related legis-
     lation;

o   to indentify options and means to strengthen countries'
     capacities and knowledge to build and to reorient the health
     services.

Methods Of The Study

Information was gathered by a Working Conference on Health Services Systems.

This working conference was held in The Hague, August 29 thru
September 3, 1982.

The programme committee was composed of Albert van der Werff, Chairman, Gary Hirsch, Co-chairman, David M. Pendreigh, Milton I. Roemer, Gunnar Wennström and Bogdan M. Kleczkowski (advisor from the side of W.H.O.). The objective of the committee was to plan a tightly structured programme intended to produce an authoritative monograph rather than a collection of independent papers. For this purpose, the conference committee invited a small group of experts to present evaluative studies on national health services systems and on subject-matters such as health systems analysis, health system manpower supply, health system financing, health system planning, health system evaluation, health system information and health system research.

The conference embraced six topics related to Health Services Systems :

1. Health Service Systems' Environment

2. Health Services System Structure

3. Health Resources Production

4. Economic Support of Health Services Systems

5. Patterns of Health Services Delivery

6. Health Services System Management

All participants had received and reviewed a copy of the formal papers before coming to the Netherlands. The process which the conference went through, started with a plenary session on the general systems approach and its relevance for the design and development of health service systems. Also health systems problems have been identified and needs for reorientation for each of the above mentioned six topics. The identified problems then were discussed in working groups which met separately to consider the topic from a special perspective. The day's topic was then concluded with a plenary session at which rapporteurs from each working group presented their conclusions to the entire conference audience. From these conclusions, a summary was prepared to be included in this volume (Section 1).

References

1) WHO, Executive Board, Study of Basic Health Services, Off. Rec. No. 206, Annex 11, WHO, Geneva, 1973

2) WHO ' Health for All ' Series No. 1 : WHO/UNICEF, Primary Health Care, Report of the International Conference at Alma-Ata, WHO, Geneva, 1978

3) WHO ' Health for All ' Series No. 3 : WHO Global Strategy for Health for All by the Year 2000, WHO, Geneva, 1981

4) Bogdan M. Kleczkowski, Milton I. Roemer and Albert van der
   Werff: <u>National Health Systems and their Reorientation</u>
   <u>towards Health for All</u> (guidance for National Decision-
   Making), WHO, Geneva, 1983.

A SYSTEMS VIEW OF HEALTH SERVICES AND THEIR REORIENTATION:

SUMMARY OF THE CONFERENCE PROCEEDINGS AND BACKGROUND PAPERS

G.B. Hirsch

Wayland, Massachusetts

United States of America

Introduction

The first three papers in this section will present the need for a systems approach, the elements of health service systems to be examined by such an approach, and the essential process entailed in applying a systems approach. This paper uses the systems frameworks presented hereafter by Dr. Roemer to describe and analyse a set of common health services problems faced by the NATO countries and to present changes recommended at the Conference to reorient health services systems and thereby deal with those problems.

The paper begins by drawing on the background papers to present common health services problems in the NATO countries. These problems are then related to common causal factors that the background papers also reveal as being shared by many of the countries. With this background, the paper then shifts from a descriptive to a normative mode and presents a set of requirements for effective health service systems that were developed by the Conference's four working groups. Desired care delivery patterns that would be found in effective health service systems and changes needed to reorient the systems to achieve these patterns are presented next. These desired patterns and changes required for reorientation are the principal products of the conferences four working groups. Required changes cover the areas of health system structure, resource production, economic support, and the management techniques such as planning needed to effect a reorientation of health services. Final sections of the paper deal with strategies for reorientation derived from a systems view and the problems and prospects of applying systems approaches to health services.

The salient finding emerging from the conference and reported in this paper is the necessity of taking multiple, well-coordinated actions in order to achieve the reorientation of health services systems. This insight, stated as such, may seem obvious. However, decision making and change, as they typically take place in the "real world", focus on only limited segments of the system at any one point in time. Attempts to control cost, increase accessibility, and re-orient care patterns generally consist of single measures that affect financing, resources, or structure, but do not co-ordinate the several actions needed simultaneously to achieve meaningful change. Many countries, for example, place constraints on hospitals to control costs while they are at the same time training increasing numbers of physicians who produce greater volumes of services and costs.

A systems approach enabled participants at the conference to begin with a common overview of health services systems and their components (resources, finances, structure, management). It then enabled them to specify desirable changes in each of the components based on the relationship of those components to the service delivery patterns requiring reorientation. The numerous individual changes that emerged from the working groups' deliberations are significant in themselves, but are really important taken together because they represent a "tapestry" of interwoven measures that must be coordinated if they are to have an impact. Use of a systems approach enabled the participants to see the need for multiple, well-coordinated actions to re-orient service delivery patterns. Though the recommendations that emerged are not sufficiently developed to represent a blue-print for the reorientation of health services systems, they at least represent an outline for such a blueprint. The final section of this paper suggests a process for using a systems approach to further elaborate these recommendations and, in doing so, develop detailed strategies for reorienting countries' health service systems.

The next section describes the common problems faced by the NATO countries and many of the underlying common causes of those problems. These common problems and causes suggest a systems framework and can be useful in many or most of the NATO countries. Developing a common framework rather than focusing on the differ-ences among countries as is often done would allow insights to be shared among countries more readily. Positive experiences with reorientation strategies in some countries can be adopted or at least evaluated more readily by others, when a common model is used.

1. Overview of Health Service Systems in the NATO countries:
   Structural Differences, Common Problems

   The papers prepared as background to the conference,
which appear in Section III of these proceedings, present the
sort of diversity one familiar with international comparisons
of health services would expect. Structural differences among
systems revealed by the papers are numerous. The health
service systems fall along a continuum that might be labelled
"degree of organization" (as Milton Roemer does in his paper).
At one end are systems such as those in the United States in
which participation is not compulsory and in which the majority
of the health insurance sector is private and free to establish
its own benefit patterns. The large majority of providers are
also private and negotiate fees with insurers in a process that
is, in most cases, free of government intervention. In the middle
of the continuum are countries such as Belgium and the Netherlands
whose health insurance schemes are compulsory but administered
through private, non-profit "sickness funds". France pays for
basic health care costs from a single social security fund, but
has numerous private sickness funds to handle co-payments. Most
primary care in these countries is delivered by private physicians and the percentage of hospitals that are publicly owned
varies among these countries. Public hospitals tend to predominate in France while having a less dominant position in the
other countries. Payment of private providers by "sickness funds"
is generally done in accordance with a set of nationally established fees or "tariffs". At the other end of the continuum are
the United Kingdom and the Scandinavian countries among others
that have health service systems in which financing occurs almost
entirely under governmental auspices. Operational management of
health services in these countries is usually delegated to the
regional and local levels while the proportions of financing
supplied by national and regional levels and by general taxation
and social insurance funds varies among the countries in this
group. Hospitals in these countries are generally government-
owned while primary care is often provided by independent practitioners.

   While each of the countries is undergoing incremental shifts
along this continuum, generally in the direction of greater
organization, some countries are undergoing more fundamental
shifts. Italy, for example, has a number of health insurance
funds that make it typical of the cluster of countries at the
middle of the continuum, but is moving toward a more unified
system of financing and care provision. Greece is undergoing a
shift toward greater reliance on the public sector for the provision of care while Turkey, with a largely nationalized system,
seems to be moving slowly in the opposite direction.

The magnitude of structural differences among the health service systems in NATO countries makes the similarities in the problems they face quite striking. Most of the countries, for example, report difficulties in controlling costs of health care, usually in proportion to the extent to which they rely on the private sector for care provision and financing. Though most offer universal entitlement to health care, there is the widespread problem of uneven accessibility of care by various population groups that is often the result of maldistributed health manpower. Many of the country papers report that these problems are in turn related to undesirable patterns of health care delivery that place heavy emphasis on physician-dominated medical care and high technology specialized care in hospitals. Such patterns exist despite calls by high-level officials for health care systems that stress primary care, prevention of illness, greater use of non-physicians to perform appropriate health care tasks, and a much more limited and clearly defined role for hospitals and the technologically-oriented health care they provide. These patterns have resulted in escalating costs driven by the availability of increasingly elaborate medical technology and increasing numbers of medical specialists whose practices revolve around the use of the technology. Since most countries seem to be limiting the total amount of money they spend on health care to a certain percentage of GNP, the result of the institutionally-oriented, physician-dominated patterns of care is to leave too little money for primary care, programs for the prevention of illness, and environmental and social services that affect health. These forces also concentrate health resources in certain cities at the expense of people in rural areas and in less prosperous regions. Other common problems frequently cited in the papers include a lack of integration between health services and related social services that do exist and a lack of coordinated services for the elderly whose needs tend to be defined as medical (even if they are not) in order to receive attention from the health services system.

The complexion of these problems varies, of course, from one country to another. In the less highly organized countries as the United States and Belgium, primary care is harder to find and many people utilize expensive, specialized resources unnecessarily for simple problems. In the United Kingdom, primary care is easier to get because of the higher status accorded the general practitioner and the per capita method of paying GPs for care, but primary and specialized hospital-oriented care are not well-coordinated. France reports its success in controlling the relative growth of the hospital sector. The Netherlands has been less successful in controlling hospital costs but is enacting new legislation to achieve tighter control. Fundamentally though, these problems have many similarities, and, it appears, common roots.

2. Common Problems, Common Causes

In tracing out the details of the various countries' health service systems, the background papers suggest that these common problems have similar causes despite what appear to be major structural differences among countries. Discussions of each of the components of health service systems outlined by Dr. Roemer's model reveal similarities.

Health Resources

In the health resources sector, for example, a common issue was the lack of coordination between the training of various types of health manpower (especially physicians) and the health care needs of the population. Training of health professionals tends to be seen as a right to be made available to as many young people as possible in each country with only limited attention to how those professionals will be utilized. Medicine as the most highly paid health profession and the specialties as the most lucrative segment of that profession attract as many people as there are places in medical schools. The resulting number of graduates increase the volume and therefore the cost of the medical care that is de-livered. Large numbers of specialists lead to an increase in the demand for hospital care as well. Training physicians, nurses, and other personnel in proportion to their numbers currently working in health service systems also helps to reinforce existing care delivery patterns and makes it difficult to reorient those patterns. Though most of the countries represented have this problem, France is one that reports some progress in controlling the growth of its physician supply. Scotland has been able to forge an effective link between its medical schools and a National Health Service that has helped to match physician production to needs.

Attempts to deal with the problem of maldistribution of health services through the production of health professionals has general-ly not been successful. The larger numbers of health professionals trained have tended to locate in areas that already have adequate supplies. Certain countries with more highly organized systems, such as the United Kingdom have had some success in controlling physician location and shifting resources toward underserved areas.

Health facilities and technology are other health resources that many countries have difficulty controlling. Eagerness to adopt new technology and to modernize obsolete facilities creates pressures that regulators have difficulty resisting. Even in systems with strong central control of capital expenditures such as the United Kingdom, private fund-raising efforts result in the purchase of new CT scanners and other equipment that the National Health Service feels obligated to accept. France, again, appears to have made some

progress in regulating capital expenditures through the strict use
of unit-to-population ratios for new technologies. This, however,
has not been done without controversy about avoidable deaths in
such areas as end-stage renal disease. Capital expenditure patterns
in many countries reinforce both existing patterns of health resources and the patterns of services that are delivered.

Economic Support

The background papers also implicate mechanisms for the economic support of health services as causes and reinforcers of undesirable care delivery patterns. In the United States for example,
third-party payment patterns that favor hospitals and nursing homes
are seen as partially responsible for a significant shift in health
expenditures toward institutional care. Other countries report similar problems and have made various attempts to gain greater control
over health expenditures as a first step in using economic support
mechanisms to reorient care delivery patterns. The Netherlands has,
for example, consolidated a large number of sickness funds into a
few large ones as a way of achieving greater control and is moving
to achieve more stringent controls over hospitals'expenditures
through the setting of tariffs and bed closings. Italy is also moving to reduce the number of sickness funds in an effort to achieve
greater control. Even in the U.S., several states are moving towards
more unified financing mechanisms that coordinate the efforts of
several different governmental and private insurers. These actions
have often been less effective than anticipated. In Belgium, for
example, there is the fear that co-payments for physician visits
introduced to control health expenditures will counteract the push
toward emphasizing primary care. Countries with more centralized
financing mechanisms such as the United Kingdom find that while it
is relatively straightforward to control total expenditures, controls on particular services are difficult because of political
pressures to preserve the clinical autonomy of physicians. The
Netherlands has found it easier to regulate hospitals than to regulate the behaviour of physicians.

Certain attempts to use economic support mechanisms to reorient
care delivery patterns have succeeded to some extent. One of the
more noteworthy attempts has been the Health Maintenance Organization
(HMO) in the U.S., an entity that combines the financing and care
delivery function for a defined population of subscribers. These
organizations have been able to emphasize primary care through the
mix of personnel they employ and have been able to provide physicians
with incentives to rely less on expensive hospital care. In the
Netherlands, there are also some controls in that specialists cannot
be paid for seeing a patient unless there has been a referral by a
general practitioner. In other countries such as Belgium and Greece
and for most patients in the U.S., there are no such controls.
Patients frequently refer themselves to specialists and are more
likely to be admitted to hospitals as a result.

In general, it appears that economic support mechanisms reinforce existing patterns of services. Because health care costs are
already so high, it is hard to justify extending benefits to services
that are less commonly covered by health insurance (e.g. screening)
even though those services could potentially help to lower costs.
Historically, health insurance has developed to cover those services
that were the most expensive and posed the greatest hardships for
patients. Not surprisingly, those services have also experienced
the most rapid growth in cost because patients were less reluctant
to obtain services covered by insurance. By directing the bulk of
insurance payments and budgets to hospitals and physicians, economic support has also helped to reinforce existing patterns of
health resources and service delivery patterns.

Health System Structure

The relationship of health system structures to care delivery
patterns was also touched on in many of the papers. Regionalization
was seen by many countries as a means of getting greater control
over health services and thereby being able to reorient delivery
patterns. Norway, Sweden, and Finland appear to have been quite
successful in achieving coherent health services by organizing the
majority of their health services at the local level, with tertiary
services and facilities planning at regional levels. Other countries have been less successful in using regionalization, apparently because they have not delegated sufficient power to regions
to effectively handle their responsibilities. The U.K. has had to
modify its regionalization scheme by eliminating one level in order
to simplify and reduce the time required for planning and decision
making.

Relationships between public and private sectors and the balance of services provided by the two are issues in the process of
being resolved by many countries. For countries with a substantial
private sector, the degree of regulation required to assure quality
and efficiency of health services and mechanisms for providing that
regulation are matters of intense public debate. The Netherlands,
for example, is in the process of implementing a broad set of regulations for private hospitals. Italy and Greece are going further
in absorbing many services now privately provided into the public
sector. Even in countries with a limited private sector such as the
U.K., there has been debate about "pay beds" for private patients
in NHS hospitals. In France, physicians at public hospitals are no
longer able to see private patients. In some countries such as Greece
and Turkey, the magnitude of care delivered by the private sector is
even difficult to estimate because much of that care is provided by
physicians who are also government employees, but have part-time
private practices.

Coordination among health services and between health and other social services is a third area touched on by many of the background papers as a troublesome aspect of structure. Though many countries such as the United States, United Kingdom and several Canadian provinces have departments of health and social or human services at the highest level, actual provision of these services is done quite separately at the local level. Clients with complex problems that have both medical and social aspects must deal with a fragmented delivery system and often receive ineffective and/or duplicative services as a result. The elderly have particular problems in dealing with this lack of coordination. The United Kingdom is an exception as far as the elderly are concerned since an emphasis on geriatrics has led to better coordination of the services they receive.

There is also a lack of coherence among the various health services in many countries. Primary care and specialized, hospital-based care are usually provided in different locations with minimal coordination in the less organized systems such as those in the United States, Canada and Belgium. People in those countries may enter the system at any point regardless of what kind of care is appropriate. Even in highly organized systems like that in the United Kingdom primary care and specialized care are administered quite separately with coordination provided more by interpersonal relationships of staff than organizational linkages.

Determinants of Health Service Demand

One can also look beyond the health service system itself to find additional common causes to these common problems. In the larger system of factors that affect a population's health and demand for health services demographic changes have had a significant impact in all countries. The proportion of elderly people has grown and with it the demand for health services including long-term care for chronic illnesses. Certain diseases to which the elderly are more susceptible have increased in prevalence and importance as a cause of death while others, such as heart disease have been declining in relative importance as causes of death. At the same time lower birth rates have reduced the demand for obstetric and pediatric services. Growth in the prevalence of alcoholism and emergence of new problems such as narcotics addiction are among a number of health problems that have placed additional demands on health services in many countries.

Medical technology has had a direct effect in many countries by increasing the intensity of services provided to patients. It has also had an indirect effect by reducing mortality rates and thereby leaving people with chronic illness alive longer to receive additional health services. Technology and specialization in medicine tend to be mutually reinforcing. This has contributed to further fragmentation of care as well as higher cost.

Values relating to health care are another source of the service delivery patterns described earlier. Most countries represented at the conference reported equity of access to care and clinical autonomy of physicians as values that profoundly affect the provision of health services. Equity of access, while essential, contributed to growing demand for services and reduced the constraint that cost would otherwise place on the utilization of care. Clinical autonomy has made it difficult to regulate health care or to resist pressures to adopt increasingly complex medical technology. In the less highly organized countries pluralism and freedom of choice among providers are also widely accepted and make it more difficult to regulate services and achieve greater cohesion. While cost control is a significant concern in almost all of the NATO countries, these values have served as obstacles to measures that could effectively control costs.

None of these values are in conflict with the ideal of health service systems reoriented toward primary care. However, their effect on the demand for the existing set of health services together with desires to limit total health expenditures in most of the countries leave little money available to fund new or expanded primary health services.

3.   Requirements for Effective Health Service Systems

With this set of common problems and causes as background,
those attending the conference began applying a systems approach
to identifying solutions and strategies for reorientation. The
first task was to specify the requirements for a truly effective
health service system. These requirements would describe a desired
end toward which reorientation could proceed.

There were four working groups, each of which examined one
of the following sources of demands on health services systems:

1. Democratic Demands for Equity.
2. Changing Demography and Spectrum of Disease.
3. Physical and Social Environment.
4. Advancement of Science/Health Technology.

The following are the principal requirements articulated by the
four groups:

Effective health services systems should:

1. Offer equal access to care for all and make special efforts to
   reach those who have not traditionally received adequate care.
   Equal access should include preventive care that is often
   viewed as a "luxury" while acute care needs remain unmet.
   Attempts to achieve equity need not produce uniformity and in-
   flexibility nor try to solve all aspects of social inequality
   through the health service system. Increased consumer participation and decentralized management of services are seen as two
   structural devices that may promote equity.

2. Look for and respond to demographic and morbidity trends in a
   timely manner. Significant trends include: 1) The growing
   proportion of the elderly with their greater need for health
   care and related social support; 2) emergence of new illness
   patterns such as sexually transmitted diseases and drug addiction and iatrogenic illness related to medical care; and 3)
   shifts in treatment modalities made possible by such developments as medications enabling the mentally ill to be treated
   in outpatient settings. Changing needs resulting from these
   trends, rather than the availability of new medical technology,
   should be the driving force for changes in the health services
   that are provided.

3. Have the flexibility to deal with uncertainty created by changing
   trends, values, and technologies. This requirement should cause
   health service systems to minimize investments in expensive,
   highly specialized resources. This type of resource cannot
   easily be adapted and instead tends to distort patterns of
   service delivery and prevent new patterns from developing.

4. Make cost-effective use of resources within the bounds of what
   society is willing to allocate. With regard to cost-effectiveness, it is especially important to assure the appropriate use
   of health technology. Though the advancement of knowledge is
   almost always a positive force, the application of that knowledge
   should be tied to evidence of a cost-effective result. Organizing
   services hierarchically with regional control of high-technology
   services may be one device for assuring appropriate utilization.

5. Be able to identify where health is affected by factors external
   to the health services system and decide when and how to affect
   the relevant service systems (e.g. nutrition, housing). Such
   problems include pollution, urban congestion, unemployment, and
   consumption of products (e.g. sigarettes) that have inherent
   health hazards. Health service systems also need the political
   skills to participate effectively in the resource allocation
   process to assure an appropriate share of the resources for health
   care.

6. Be able to evaluate its own performance in meeting the above requirements and reorient themselves when necessary. The ability
   to reorient requires a commitment to necessary changes even
   when they conflict with entrenched interests and possibly to
   spend additional money on new services while existing ones are
   being phased out.

Discussions of the working groups and the background papers on the
various countries suggested that a good deal of reorientation is
needed for most countries' health systems to meet these criteria
of effectiveness. A sequence of sessions ensued in which the conference participants described service delivery patterns that
would exist and other health system changes necessary if such a
reorientation is to take place. The next section presents the results of these sessions.

4.   Initiatives for Reorientation: <u>Findings of the Working Groups</u>

After defining requirements for effective health service
systems, the remainder of the conference was devoted to describing
the components of systems that meet these requirements. A sequence
of working group sessions focused on each of the components, con-
tained in Dr. Roemer's model: health system structure, health resources, economic support, service delivery patterns, and management. Each of the four working groups covered one topic within each
component. A plenary session following each set of working group
sessions allowed the groups to share their findings and develop
additional ideas on the topics being discussed.

The purpose of these working group and plenary sessions was
to respond to the requirements described in the last section and
problems alluded to in the country background papers by describing
how health service systems could be reoriented. This paper first
presents findings about desirable service delivery patterns even
though, in reality, these were discussed later in the week. Discussing these first will provide a better context in which to then
present findings about structure, resources, and economic support.
Discussions of structure, resources and economic support are then
tied back to what is required to achieve these desirable services
delivery patterns. The section ends with a discussion of the management techniques that are needed to implement a reorientation.

Systems lessons arise in the course of examining each of the
five components of health service systems. Controllability, hierarchical structures, measurement and evaluation, definition of
goals and objectives, and integration are systems concepts that
come up repeatedly. However, as suggested earlier, the most striking
contribution that systems approaches can make to the reorientation
of health services is in helping planners and managers come up with
a coordinated set of actions that combine needed changes in structure, resource development, and mechanisms of economic support.
Strategies for reorientation that arise from systems insights developed at the conference are presented in the next major section
(5.)

Desirable Service Delivery Patterns

The working groups addressed four topics in defining desirable
service delivery patterns: 1) the preventive/curative care mix,
2) integration of care, 3) coherence of services, and 4) primary
health care.

The group working on the <u>preventive/curative care mix</u> first
defined the types of preventive care. Health promotion, one type,
was seen as the removal of environmental and behavioural risk

factors. The other type, disease prevention, is devoted to the prevention or postponement of specific diseases. Primary preventive measures such as immunization keep certain diseases from occuring at all while secondary prevention emphasises early detection and control of specific conditions such as high blood pressure. Tertiary prevention places its emphasis on control of existing chronic conditions and prevention of complications (e.g. as in the case of diabetes, heart disease).

The group recommended encouragement of health promotion through the advocacy of healthful practices to the general public and health providers, implementation of environmental controls (e.g. on air and water pollution), elimination of occupational hazards, and provision of positive incentives such as reduced life insurance premiums for non-smokers. Caveats mentioned by the group include a lack of proven cost-effectiveness for many forms of health promotion and possible adverse economic effects resulting from limits placed on certain industries.

Several approaches were recommended to encourage preventive measures. Persuading the various medical specialties, through continuing education and change in medical school curricula was seen as a key step. Changing health insurance benefits and reimbursement schemes to shift emphasis to preventive care from high-technology curative care was seen as another necessary step. In the absence of specific financial incentives for preventive care, physicians on salary or capitation might be expected to provide more preventive care than their counterparts paid on a fee-for-service basis. Payment mechanisms should also make provisions for reimbursement of preventively oriented professionals such as health educators and exercise physiologists and for activities that train patients in self-care. Finally formation of a nationally prestigious organization perhaps of a mixed public/private character, may be necessary to give health promotion and prevention sufficient priority in the allocation of a nation's health care resources.

Primary care is the setting in which many preventive activities occur. It was also seen by the working group concentrating on it as a source both of first contact care (i.e. a sort of "gate-keeper" role) and continuous care of chronic conditions when specialist or hospital care is not called for. It is also a source of care for health-related social and emotional problems that can be amenable to counseling. Though there seemed to be a lack of adequate primary care in most countries, the group felt that the reasons differ from one country to another. In some, it is a lack of general practitioners while in others it is overly-easy access to specialists.

Health centers using a team approach among several professions such as those found in the United Kingdom and Finland offer a po-

tential mechanism for overcoming these difficulties, especially a
shortage of general practitioners. Some in the working group felt
that decentralization and regional management could help to promote primary care. Others however, were not so certain of this.
Cost controls were seen as another device that could cause a shift
from specialized to primary care, but it is unclear that such a
shift could actually lead to cost savings. There might even be
higher costs as a result of screening done in primary care settings
that reveals new cases requiring treatment. Continuity of care for
chronic illnesses made possible by primary care settings could
offset some of these higher costs by achieving the objectives of
secondary and tertiary prevention discussed earlier.

A third group concerned with the issue of <u>cohesion</u> of health
services also emphasized the importance of the single, well-defined
entry point that primary care sites can provide. Its members felt,
however, that what is gained in cohesion by limiting entry points
would be lost to some extent in reduced freedom of choice. The
group also felt that increased professionalism and specialization
tend to reduce cohesion and need to be limited if cohesion is to
be achieved.

The fourth working group dealt with another desirable pattern
for service delivery, the <u>integration</u> of care among various providers including those outside the health services system whose
activities nonetheless affect a population's health. The group found
integration a quality that can range along a continuum from
loose cooperation to well-structured coordination, and principally
involves the actual delivery of care (as opposed to financing,
management, etc.). Groups of people requiring treatment for chronic
illness, the disabled and the elderly were seen as those who could
most benefit from integration of care. These people tend to have
problems that are cumulative, interdependent, long-term, and only
partly medically defined or treated. For these people, integration
of care should help to: provide ease of access to services, deliver
effective combinations of services needed to deal with multi-
faceted problems, reduce cost as a result of eliminating duplication of services, avoid provision of excessive levels of care,
enhance self-reliance by providing support systems to compensate
for partial handicaps, and provide clearer definition of client
groups for political/advocacy purposes.

Integration could, however, produce negative side-effects.
One could be stigmatization of the target group. Another one is
the possibility that integration could concentrate additional power
in the hands of health service subsystems that already have a
great deal, simply because they seem to be logical points for lo-
cating integrative activities. Introducing integration into poorly
organized health service systems could also complicate rather than
solve problems.

The group suggested that it would be more appropriate to develop integrated responses to health problems than new integrating
agencies. A desire for integration should instead result in better
coordination among existing agencies. Such coordination should include both public and private agencies, cross professional boundaries, and be voluntary, but with the obligations of each agency
clearly defined.

As indicated in the background papers discussed earlier, the
desirable health service delivery patterns described above are not
heavily emphasized by most countries. What initiatives in the other
health system components are needed to put greater emphasis on
prevention, health promotion, and integration of services? The
following sections provide some ideas.

Health System Structure

The working groups dealt with four aspects of structure:
1) public sector, 2) private sector, 3) regionalization, and 4)
consumer participation.

The groups focusing on the public and private sectors both
began by defining and characterizing each sector before describing
desirable structural attributes. The private sector was depicted
as the part of the system that responds more readily to economic
incentives. It can react more readily to changing demands once
those demands have been clearly recognized and a means of payment
has been provided. The public sector, on the other hand, responds
more readily to broader, more complicated social problems and can
take risks that the private sector avoids. This is because the
public sector is able to draw on general tax revenues while the
private sector usually relies on revenues derived from service
delivery for most of its support.

In terms of recommendations, the groups found that the
appropriate level of private sector participation in a country
should be accompanied by suitable forms of regulation to assure
that health needs are met since this may not be guaranteed by
economic incentives alone. Regulation should include sufficient
accountability to assure the efficient use of resources and adequate quality of care. It should however, be based on broad objectives rather than detailed regulations to avoid bureaucratizing
the private sector. Payment mechanisms also need to be better
oriented to health needs and cost-effectiveness to motivate an
appropriate response from the private sector. Regulations such as
licensing of health personnel should allow flexible private
sector responses to changing needs while, at the same time, protecting the public and assuring quality.

Public/private partnerships with clear definition of each

sector's obligations are often the most effective way of meeting
particular health needs. At the national level, Health Councils
representing both the public and private sector providers can
provide a useful forum in which comprehensive policies can be
hammered out and commitments made. In addition to policy development at the national level, public-private coordination at the
regional level is essential for strategic planning and resource
allocation and at the local level for the organization and delivery of services. Philosophy about the respective roles of the
public and private sectors and the balance between the two should
be reexamined periodically to assure that they continue to be
consistent with national goals and values.

Regionalization was seen by the working group discussing it
as a countinuum along which there are several distinct models
representing differing degrees of autonomy for the regions. True
regional autonomy (rather than what the group called a "good
behaviour" model) requires constitutional protection to keep
autonomy from being arbitrarily rescinded, a political forum at
the regional level in which to debate policy, law-making competence, and the power to raise the funds at the regional level to
pay for health services.

Regionalization was seen to provide an opportunity for
"rational" allocations of functions with each one performed at the
most appropriate levels. Primary health serviees would best be
provided and managed at a local (city or town) level while an
entity such as a county could be the level at which a full range
of the usual hospital specialties would be provided. Superspecialty
hospitals and medical teaching programs would serve an entire
region while broad policy guidelines and goals and objectives for
health services would be set at the national level.

Making regionalization work in a manner such as this presents
the planner of a regionalized system with several challenges. The
principal challenge is to avoid or ameliorate the tension between
the regions' desire for greater autonomy and the central authority's
perceived need for control. This requires that directives from the
central authority be neither too specific and thereby erode the
autonomy of the regions nor too vague as to lack credibility as
genuine guidance. Another challenge is to assure sufficient cooperation between public and private providers and between the
health and social welfare sectors at the local and regional levels
without the central authority's detailed "orders" on how to do so.

Consumer participation was seen as an aspect of health service
system structure that is becoming increasingly important for several
reasons. These include a response to perceived public discontent
with health services and resulting demands for change, implementation
of government policies to balance the decision-making powers of

health service providers and managers, and to increase consumers' involvement in the solution of health problems (including caring for themselves and their families).

A number of issues were raised by the working group concerned with consumer participation. One was the need to assure that consumers' desires are adequately understood, through surveys and other means, rather than simply assuming that their representatives accurately reflect broad consumer opinion. Poorly educated members of minority groups pose special problems for consumer participation and these need to be remedied. All consumer participants need the proper knowledge base in order to be involved in a meaningful way. Some control over expenditures would also make participation more meaningful. Adequate knowledge is needed to avoid pressures for funding projects that have little medical value. The benefits of increased participation would be better health education for the population as well as improved decision making.

A review of the working groups' findings suggest several ways in which structural solutions can contribute to reorientation. One is that a coordinated strategy involving the public and private sectors may have a good chance of succeeding. The public sector can sponsor programs in areas where the efficacy of prevention and health promotion is unclear. Once efficacy has been demonstrated, a means of payment can be identified and the private sector given responsibility for widespread dissemination of the preventive techniques. Well-designed regulatory mechanisms would be necessary to assure that the private sector is performing its role appropriately without reducing that sector's inherently greater flexibility to respond.

Integration can also be accomplished through a public/private partnership if the public sector takes responsibility for identifying target groups and their multiple needs and contracting with private agencies to actually meet those needs.

Regionalization can assist in the reorientation of care delivery patterns in several ways. For one thing, working at the regional level reduces the scale of problems that must be dealt with and the number of actors involved. Regional strategies can be tailored to fit regional needs rather than having a single national strategy that badly serves many of the regions. Working at the regional level also enables better control of new programs that develop to assure they are meeting their objectives. Allowing variation among regions can also provide a natural laboratory in which alternative approaches are tested. Though some regions may fail initially, learning will take place and new programs may eventually succeed. If a single, monolithic approach is tried at the national level and fails, widespread disillusionment may occur and prevent other approaches from being implemented. For regionalization to

produce these benefits, however, it appears essential for the
regions to have sufficient (i.e. real) autonomy and national
governments to concentrate on broad policy formulation rather
than detailed control.

Consumer participation can aid in reorientation by monitoring
providers to make certain they are genuinely committed to health
promotion and prevention, especially since many providers derive
the bulk of their revenues from curative activities. Participation,
as was indicated earlier, could itself be a vehicle for health
education and promotion activities.

Health Resources Production

The discussion of health resources by the working groups
touched on four areas: 1) health manpower, 2) health training and
education, 3) health facilities, equipment, and supplies, and 4)
health technology.

Concerns discussed by the groups focused on health manpower,
and health training and education overlapped to a large extent.
Both groups stressed the need to consider the full range of personnel
engaged in the provision of health services rather than merely focusing on physicians. The emphasis should be on the identification of
health needs that can be met by the various types of personnel and
the provision of numbers of each type in accordance with the health
needs of the population. Existing patterns and ratios of manpower-to-
population should not be the sole basis for manpower planning as
they often have been in the past.

With regard to physicians, one group discussed whether it would
be more desirable to broaden the physician's training to include
means of dealing with social needs (e.g. counseling) or to instead
narrow training to deal with strictly biomedical concerns. After
much discussion, the group felt that physicians represented too expensive a resource to be applied to needs that could be handled as
well or better by lower-paid practitioners. There is also a danger,
if physicians'roles are defined too broadly, of mislabeling various
health needs as "medical" and inappropriately requiring a doctor's
care.

Although a narrower role for physicians was seen as desirable,
both groups cautioned against the inflexible segmentation of health
care tasks into a number of narrowly defined professions. Such segmentation leads to the inefficient utilization of health manpower as
well as the fragmentation of care. A single planning scheme for all
of the health professions is needed to arrive at the appropriate
degree of segmentation and numbers of each profession that are
needed. The region and its population may be the most appropriate

level at which to do health manpower planning. As suggested earlier,
the region's needs rather than norms based on current patterns
ought to be the basis for such planning.

Manpower planning must also be done with sufficient attention
to systems of remuneration. Higher remuneration causes students to
want to become physicians and other highly paid professionals.
This, however, needs to be reconciled with the population's needs
for particular health services. Remuneration may also be a means
for assuring adequate numbers of personnel in underserved areas.
Salary and capitation rather than fee-for-service, are forms of remuneration that offer the greatest potential for controlling both
the numbers and geographic location of health professionals.

Both groups agreed that the most important means of achieving
control over health manpower supply and distribution is close coordination between the health services system in a given country
and its universities and other institutions that train health manpower. Scotland was again pointed out as a place that had achieved
a reasonable degree of success in coordinating training and health
services, especially in the case of physicians. Even when there is
good coordination within a country, migration complicates matters.
The alternative means for matching manpower supply to needs mentioned was to open educational programs to everyone who wants to
enter (as in Belgium) and allow market forces to provide the control.
For this method to work, however, finances must be constrained or
there is a good chance that the volume of health services consumed
will simply rise to absorb the larger supply.

The quality of health manpower was also emphasized as an important issue. Basing assessments of quality on competency rather
than knowledge base was seen as more desirable, but is often
thwarted by licensing regulations tied to educational curricula.
The ideal would be a lifelong process of self-assessment and continuing education based on competency.

The education of consumers to take better care of themselves
was seen as an additional way to produce a health resource.

The group concentrating on <u>health facilities, equipment, and
supplies</u> focused their attention on potential reasons for inappropriate decisions about needed resources. Reasons for poor decisions
might include the use of poor quality information, use of standards
for monitoring system performance based on planners' values rather
than those of the general public, a lack of agreement among separate
agencies that affect resource decisions, and conflicting interests
that must be resolved (e.g. a desire for high occupancy levels to
permit efficient utilization of hospitals and a conflicting desire
to keep people out of hospitals who can be adequately treated at
home).

A major point discussed by the group was the need to co-ordinate health service system's actions with those of related systems whose actions also impinge on health (e.g. the traffic safety system). Techniques for achieving such coordination include finding mutual advantage between systems, bargaining if there are favors or resources that can be exchanged, and advocacy. The implicit idea that comes out of the group's work is that health needs do not necessarily translate into health facilities and equipment. An effectively functioning health service system would first see whether negotiation with other sectors could help to meet those needs before investing in expensive health facilities and equipment that can only be used for limited purposes. For example, improvements in traffic safety may do more to reduce fatalities and disability than sizeable investments in facilities for trauma care. Avoidance of large investments in facilities and equipment will also permit greater flexibility in reacting to changing demographic trends and treatment methods.

The group discussing health technology as another type of resource first focused on its benefits. These include saving life or reducing the burden of illness, reducing cost or increasing cost-effectiveness, increasing diagnostic accuracy, and making procedures safer. The group felt that the problems were not so much with the technology itself, but with its applications. Inappropriate use of technology also relates to the environment in which it is used. Some technologies may be unsafe if used outside of the hospital while others are perfectly safe for ambulatory patients and would be misused if they led to unnecessary hospitalization. Certain procedures such as tonsillectomies may themselves be unnecessary in many cases.

To avoid abuse, technology requires careful evaluation of its efficacy and appropriate uses. Careful trials would be needed to both assess efficacy and determine the likelihood of possible harmful side effects. In addition to this sort of evaluation, several other measures would help to assure the appropriate use of technology. These include education of patients and provision of accurate instructions on use (e.g. of a new medication), education of health personnel who will use the technology and change in relevant curricula, and regulation protecting patients from abuses that may be motivated by what physicians and hospitals perceive as opportunities to generate additional income.

These findings again provide several insights about pre-requisites for reorientation. The first is that planning of changes, manpower, facilities, and other health resources needed for re-orientation should be tied to demonstrated health needs of the population rather than existing resource patterns. Preventive and health promotion tasks need to be clearly spelled out and re-

lated to the kinds of health professionals best equipped to
perform those tasks. While there is likely to be some benefit
from designating and training new types of professionals for prevention and health promotion, it is important to avoid the fragmentation that could result from over-specialization and the
development of rigid boundaries between professions. Reorientation
or adoption of existing professions should, perhaps, be explored
before new professions are created. Certain health professionals
should also be trained to function effectively as integrators
of care. Physicians should have some role in prevention, health
promotion, and integration of care, but not a central one since
they are a more expensive resource than is generally needed for
such activities. Coordination between universities and other
training institutions and planners is essential if new patterns
of resource availability are to emerge.

Reorientation will also be aided by a more critical evaluation
of new medical technology and perceived needs for health facilities.
New technologies should be looked at against the broad range of
possible investments in health resources and their relative benefits. Neonatal intensive care units, for example, can save seriously
ill infants with heroic measures, but absorb resources that could
be devoted to better prenatal care that could benefit many more
children and potentially prevent those problems. Similarly, open
heart surgery can extend the lives of patients with heart disease,
but competes with potentially more effective preventive programs
for resources. Concentrating resources on lower technology, preventively-oriented services should increase the cost-effectiveness
of health care while permitting greater flexibility in adapting to
future trends. Limiting the high technology curative resources to
a few tertiary centers in each country can help to assure that they
are used appropriately, yet are available to those patients who
really need them.

There is also a need to look outside the health service system
when seeking resources to meeting preventively oriented objectives.
Other systems, such as the one concerned with traffic safety (roads,
police, laws, automobile design, insurance) may provide more cost-
effective solutions to health problems (traffic fatalities) than
investing the same amount of money in health resources.

Economic Support of Health Services

The four working groups each dealt with one of the following
areas related to economic support: 1) input control, 2) output
control, 3) budgetting and 4) financing.

A common thread running through the discussions of all four
working groups was the control of the volume and pattern of health
services by means of financial mechanisms. From a systems stand-

point, one has to ask whether the health services system is purposeful before one can ask if control is possible by any means.
Members of at least one group were split on this issue. Some felt
that health service systems merely reflect the net effects of
actions by providers, patients, financers, government, and others
rather than a set of purposeful control processes working toward
specific common objectives. It was felt that a lack of purposeful-
ness and specific objectives make it easier for politicians as
well as the general public to avoid coming to terms with the
inherent conflicts among cost, access, and quality. Even those who
felt that health service systems could and should be purposeful,
acknowledged the difficulty of controlling when objectives change
over time. There is also a lack of adequate measurement and evaluation systems and accurate, unbiased data for use in those systems.

Economic inputs to health services can be controlled by public,
private, or mixed measures and institutions. Control mechanisms
can include legal or political means and entail bargaining with
providers, price controls, licensing of providers, and certification
of new health facilities among others. Application of the appropriate input control mechanism will depend on whether funds come
from national or local taxation, compulsory insurance, private payment by patient or employer, or charity.

There are several possible difficulties that can be encoun-
tered with input control. One is that in countries where there are
a number of methods of payment for health services, a lack of coordination among agencies and organizations that pay for care makes
it hard to achieve control of inputs. A second difficulty is that
the efforts of those who control the finances in many countries
are not coordinated with those who actually provide the services.
Because revenue derived from health services delivery can be in-
vested in additional inputs (personnel, facilities), control of
inputs is dispersed among financing entities and health service
providers. Without coordination, inputs will often flow to institutions and organizations that already have the greatest economic
strenght, regardless of the desire of financing agencies to limit
those inputs. A third difficulty is that there are considerations
other than financial such as status, competition, and perceived
demands for health care that affect input decisions (e.g. the decision of a hospital to expand as an expression of community pride).
Attempts to control inputs solely through financial means are therefore not likely to succeed.

Control of the outputs of health services is fraught with the
same difficulties of dispersion of control among many agencies and
organizations. There are also some additional problems. As mentioned
earlier, a lack of clear objectives is certainly one of them. If
objectives are developed they should specify desired outcomes
(e.g. longevity, levels of functioning) rather than solely focusing

on outputs such as visits and hospital admission. Objectives
should deal explicitly with the tradeoffs among cost, access, and
quality and specify desired levels of cost effectiveness rather
than merely limits on cost. Acceptability of services and
consumer satisfaction also ought to be reflected among the objectives.

Measurement was the other area in which output control problems
were identified. Control requires measurement. Because control is
in many countries scattered among a number of entities, the measurement systems that do exist deal with fragments and do not give a
picture of how the overall health services system is performing.
Overall performance measures and measurement systems should
reflect desired outcomes (if they can be agreed upon) and must
transcend the boundaries of the individual agencies that provide,
finance, regulate, and consume health service. The data supplied to
such measurement systems should, if possible, be accurate and avoid
bias and cover areas relevant to decision making. Data collection
should be stable in order to allow longitudinal data to accrue on
a given set of variables rather than changing in response to each
new issue that is raised.

Related to the earlier discussions about regionalization, the
group discussing output control felt that a reduction in scale (to
regional and local levels) might be essential for achieving effective control. Errors made at those levels would also produce less
damage and more innovation might therefore be possible. Political
discussions needed to define clear objectives might also be more
meaningful at those levels, closer to the actual provision of
services, rather than on a national scale.

The use of budgets as control mechanisms and respective roles
of central authorities and regions appear to vary depending on the
degree of centralization of a particular country's health service
system. In highly centralized countries such as Great Britain and
Italy, budgets are used to specify how regions will spend centrally
raised and allocated funds while in countries that have opted for
decentralization, such as Sweden, centrally-approved budgets are
guides to how funds raised at the regional level will be spent.
Both approaches can work if properly administered. However, this
requires a dialogue between the two levels to avoid an overly
"mechanical" approach at the national level that ignores regional
needs or a responsiveness to political constituencies at the regional level that leads to excessive costs.

Different financing mechanisms were seen by the final working
group as being capable of differing degrees and types of control.
Direct payment is more capable of control and adaptation to changing
demands than taxation based systems, for most existing health
services, but taxation is perhaps a better source of funds for

catastrophic illness care where patients have little choice and preventive services and where the utility of receiving such services is not clear. Taxation is also a more appropriate source of funds for developing services to meet future needs and may allow greater equity in the receipt of services among socio-economic groups. The lower degree of control possible with taxation is due largely to the need for consensus for changes under a taxation-based system and the length of time it often takes for consensus to develop. Maintaining control over existing services while developing new services to meet broader health goals probably requires a combination of the two methods of payment.

Findings in the economic support area suggest several additional prerequisites for reorientation. One is that economic support mechanisms give sufficient importance to prevention and health promotion. For this to have an impact, financial controls used to promote reorientation must be based on explicit objectives regarding these services that have been agreed upon by the various funders, providers, regulators, and consumers of care in a particular country. Efforts to implement the controls should also be coordinated among these parties lest they become adversaries and seek to undermine the process. Measurement systems that reflect progress in reorientation and impacts of reoriented services on the population's health also need to be developed. Funds derived from taxation and paid to providers on salary or capitation are probably the most appropriate form of economic support for developing preventive services and health promotion. As these services become a more prominent aspect of health service delivery patterns, a shift to support by private funding on a fee-for-service basis may be practical.

Economic support mechanisms not only affect service delivery patterns directly, but also affect health resource production which, in turn, also has long-term effects on service delivery patterns. The heavy emphasis of most payment mechanisms on curative, medically oriented services causes resources to be heavily skewed in this direction (e.g. specialized physicians, hospitals, expensive medical equipment). Because there is concern about limiting overall expenditures for health care, there is usually very little left for preventive services and health promotion. It is therefore difficult to get people to enter preventively-oriented professions when there is no guarantee that there will be funds to pay them. Resource patterns are not likely to change without sufficient attention to payment mechanisms.

Management

The final set of working group sessions and plenary sessions dealt with the management techniques that would be instrumental in helping to reorient health services. Four areas were examined: 1) planning, 2) information systems, 3) monitoring and evaluation,

and 4) research. Since these areas and others are dealt with in
some detail in a series of background papers that come later in
this report (see Section II), they will be touched on only briefly
in this section.

Planning, according to the first working group, cannot be
entirely comprehensive in considering all factors and must therefore interface with the work done by other disciplines to be
effective. The group felt that good planning has several important
characteristics: it is anticipatory, flexible, coordinates actions
that occur at different times and places, comprehensive to the
appropriate extent, systematic, integrative, participatory, and
responsive to professionals, consumers, and government. Orienting
planning toward outcomes was seen as important, but difficult
because it is often hard to demonstrate a relationship between
planned programs and desired outcomes. Planning also needs to be
coordinated among different hierarchical levels, but this is not
always done.

Planning has an essential role in the reorientation of health
services because, as alluded to in the previous pages, a number of
changes must occur in a carefully thought-out sequence over time.
Changes need to occur in structure, resources, and financing if new
service patterns are to develop, but each of these changes takes
time and are interdependent with each other. Plans need to consider
all of these activities, but cannot consider each of them in great
detail. As the working group suggests, planners must involve the
other key actors such as providers, regulators, and consumers if
planning is to be successful.

The role of management information systems should produce
neither too much nor too little information. Who requires information and what information is needed are questions that require
careful examination. If they are not, the "information explosion"
made possible by computers can lead to excessive information
being available with too little of that information being useful
for management purposes. The working group concerned with information systems felt that information should be readily available
to all "problem owners" in the health care field and that information be objective in order to help support a climate of trust.
Available information should cover health problems, consumer demand
and needs, resources, service volumes, costs, and impact on health
status in order to provide a sufficiently complete picture.

As indicated earlier, information systems are an essential
part of the control processes that are needed to assure that reorientation takes place once the new service delivery patterns
have been planned. Access of information to providers, regulators,
consumers, and others is necessary since all of those parties must
play a role in a successful reorientation. Information should cover

health status and the population's needs as well as more generally available data on resources and volumes of services delivered in order to determine what progress is being made and what further shifts in resources are necessary.

Monitoring and evaluation were seen as the collection of data about ongoing progresses and comparison of those data to particular standards. Monitoring should track accomplishment of desired standards at three different levels: 1) long-term goals expressed in general qualitative terms (e.g."Health for all in the year 2000"), 2) objectives for particular activities that are desired, and 3) target levels of activity and outcome to be achieved by a given point in time. At the third level, accomplishment of targets, monitoring should be concerned with both the attainment of health objectives and the efficiency with which resources are used.

Evaluation can deal not only with the attainment of a particular standard, but with appropriateness of the standard itself. As time goes by, environmental changes and progress toward attaining objectives may suggest revisions in standards. Information collected by monitoring processes may also suggest changes in the assumptions upon which objectives are based (e.g. that a particular preventive measure is not as effective as first thought) and should be evaluated for this purpose as well.

Reorientation is not likely to be a smooth process in which plans are readily implemented. Instead, implementation will be a rough process characterized by movement ahead and sliding backward. Much learning will take place as reorientation proceeds. Monitoring and evaluation systems have a critical role to play in the learning process.

Research is also an important part of the collective learning process needed to make reorientation work. In a systems framework, research is subservient to other management processes and provides new information for decision making. The raising of research questions according to the working group concerned with research, should again be open to all of the groups involved in health services. Research should be carried out in a manner appropriate to the task. For example, research done to support pending policy decisions may have to be done with less refinement than pure scientific work in the interest of producing a timely result. In order to be useful, research will need to deal with some of the more difficult, basic questions about the nature of health and health services and cultural influences on the development of useful health services. Research funding should come from multiple sources rather than principally government in order to eliminate the possibility and appearance of bias.

Dissemination of research results should be done in a manner that makes them available to all interested actors rather than being limited to scientific journals read primarily by other scientists. At one level, research results need to be in a form that can be used by policy makers. At a second level, the research should be understandable by the general public and contribute to a healthy public debate on health services. Reorientation of health services ultimately depends on broad public understanding of the need and support for its objectives.

The next section summarizes the broad strategies for reorientation that can be derived from a systems view of health services.

5.   Strategies of Reorientation: Lessons from a Systems Approach

What have we learned from applying a systems approach to
health services? One thing certainly is that by forcing ourselves
to examine the entire system, we have identified a great many
pieces of the system that determine health service delivery patterns. More importantly we have identified the interconnections
among the pieces. New health services, for example cannot be provided without appropriately trained personnel. It is difficult to
attract people into the training programs that produce those personnel without assurances that payment mechanisms will cover those
new services. It may be very hard to change financing policies if
the benefits to be derived from the new services are not clear.
Trying to change any one of these components alone without adequate
attention to its relationship to the others is not likely to
change the pattern of services that are delivered.

Unfortunately, in most countries, changes in single components
are the usual course of health policy making. Manpower policies,
cost controls on hospitals, the equity of various financing mechanisms, and the hierarchical structure of health services become
the issues, ends in themselves. These issues are not usually seen
in the larger context of health service systems and health of the
population. Changes are often made to relieve immediate public
pressures rather than as part of a larger strategy to improve health.
It is no wonder that the problems described in the various country
papers are so similar and so widespread.

Perhaps the most compelling "systems lesson" coming out of the
Conference is, as mentioned earlier, that strategies for reorientation must be based on a coordinated approach with attention to all
components of health service systems. Demonstration programs that
introduce new services in a limited number of settings are generally
not enough to ensure widespread adoption. The health system infrastructure must be prepared to support and nurture new programs with
resources, funds, direction, and appropriate structural adaptions.
Heavy commitments of resources and funds to existing service delivery
patterns leave little for new programs unless deliberate strategies
furnish what is needed.

Coordinated strategies require an understanding of the interdependencies among health system components over time. Certain
changes take longer to achieve than others. Long-term and short-
term changes need to be coordinated. Short-term measures should concentrate on impacts that build political support for and are needed
to sustain the long-term measures. The short-term measures should
also provide the necessary foundations for long-term changes.

A successful reorientation strategy will also entail the acceptance of a several year cost "bulge" as costs of new services

are superimposed over those existing service patterns. Thus acceptance will be difficult to achieve because of the already high costs of health services and pressures to reduce or at least contain health expenditures. Until enough time goes by and the new preventively oriented services have had time to affect health status, it is not reasonable to expect lowered costs. In fact, as discussed earlier, new preventive measures such as screening can identify previously undetected conditions that require medical care and thereby contribute further to increased costs. A longer-term effect of preventive measures is likely to be higher costs that result from longer life-spans and a larger elderly population that requires more services. Cost-containment measures are an important adjunct to a reorientation strategy because they can help to reduce the size of the short-term bulge as well as higher long-term levels of cost. However, costs are likely to be higher temporarily and, perhaps, permanently regardless of what can be done to contain cost. Creating an acceptance of this fact is an essential part of any reorientation strategy.

An important aspect of the cost-containment element will be to avoid actions that build in higher costs unnecessarily. This will require deemphasizing high-technology, high cost approaches to health problems and putting resources into services with lower capital investments and lower degrees of specialization. Uncertainties about how health service availability and the broad array of environmental and social factors will affect health and the demand for health services require the most flexible approach possible to developing services and resources. Critical evaluation of new medical technologies will be an important part of avoiding these built-in costs.

Systems approaches stress the proper definition of boundaries as a first step. Discussions at the Conference suggested that health service system boundaries have been drawn too narrowly and rigidly. Reorientation will require that the system boundaries be relaxed somewhat and that components of related systems be included in solutions to health problems. This is especially important in developing health initiatives for the elderly. Older people need many support services that affect their health (e.g. nutrition, housing, transportation), but that are not necessarily provided by health service systems. Integration of services including those provided by other systems is therefore needed.

Controllability was another aspect of systems that figured importantly in discussions at the Conference. The consensus seemed to be that the changes needed for reorientation could best be put into motion and controlled on a smaller scale, such as that provided by a region, than at the national level. Government, it appears, should not try to control too much, lest it be overwhelmed

and lose control altogether. In addition to avoiding overly centralized control of health services, government should do what it does
best and leave to the private sector the provision of well-estab-
lished services for which funding has been arranged. Governments
most valuable role is to identify needs and initiate new services
or enhance those at an early stage of development. Regulation of
the private sector should assure adequate quality control and equity
of access without overly bureaucratizing either the public or
private sectors.

Control will also require better data and the management information systems to provide the data. Health status measures as well
as the more typical data on health services and resources should be
available to measure the progress achieved by reorientation and
its impact on a population's health.

Reorientation will depend on countries' ability to turn broad
national desires for better health at reasonable cost into specific
actions. Coalitions will have to be built to support these actions
among such groups as business, labor unions, and consumers that
have not traditionally had a strong health service decision making
role in many countries. Existing health service systems are so
large that they have a great deal of inertia. Specific objectives
and actions, a great deal of coordination and planning, and political
support are necessary if that inertia is to be overcome and reorientation accomplished. Systems approaches can help to provide the
frameworks in which to identify the full set of changes that need
to occur for a successful reorientation and to build a consensus
around those changes.

The next section describes some prerequisites for effective
use of a systems approach.

6.   Systems Approaches and Health Services

How can systems approaches be most useful in reorienting health service systems? The previous section presented some broad conclusions about desirable strategies for reorientation derived from a systems view. These results of the Conference are a good starting point, but need to be taken further if they are to be of practical use. Because systems approaches are equally applicable on any scale, they can also be quite useful for the detailed design of the reorientation strategy for each country and for the many separate initiatives that will make up a country's reorientation program. The findings of the working groups at this Conference suggest many of the initiatives that will have to be pursued.

Each separate initiative such as an effort to change the composition of a country's health manpower supply will have its own systems aspects. In that case, coordination of the several groups that train and regulate health manpower, proper sequencing of long- and short-term activities, and proper attention to control processes are all as essential at that level as they are for over-all strategy development. Systems approaches can then be of further value for integrating those separate initiatives into a coherent overall strategy for reorientation.

What are the prerequisites for the effective use of systems approaches in health services? The Conference itself, if viewed as a microcosm of the health service field provided some useful les-sons. Though all of the participants were at the very least sympa-thetic to and interested in systems approaches, there was great variation in their expertise in such techniques. Experience at the Conference suggested that a first prerequisite might be to educate all of those involved about the potential benefits of systems approaches and some simple, specific techniques for carrying them out. Without the common framework that this education would pro-vide, there is a danger of confusion about what is meant by "doing systems work" and about which of many systems techniques is being used.

Involvement of a full range of people with different roles in health services is essential for two reasons. One is that these people together possess a greater deal of knowledge that is needed to assure that systems approaches are both comprehensive and realistic. The other is that participation in the systems applica-tion will help to build commitment to using its results in real-world policy making. Learning from systems approaches comes largely from the doing. Results of systems efforts can be "packaged" and communicated to others, but less effectively than if they have been involved all along. It is therefore important to involve some of a country's key decision makers in a systems application in

order to assure that the results will be used at the highest
levels at which policy is formulated.

The relationship between systems approaches and empirical
research also requires some attention. The two are not mutually
exclusive as is often the impression. They are, instead, complementary. Data about the real world provide much of the raw
material from which systems frameworks can be assembled and a
baseline against which results can be evaluated. Systems frameworks, on the other hand, can help to identify the critical
research questions to which answers are needed for management,
decision making and the development of reorientation of strategies.

A final point worth making is that the use of systems approaches in several countries for reorientation of health services
can be mutually reinforcing. Multiple approaches can be evaluated
and compared. The existence of common problems with common causes
suggests that systems approaches pursued in several countries
will themselves have much in common. Sharing insights among those
countries will allow all to benefit as well as providing a common
framework in which to carry on discussions of health service systems and their reorientation.

The work done at this Conference has provided a valuable
starting point if not a completed blueprint for reorientation of
health services. Further use of systems approaches in a number of
countries can contribute to the completion of suitable blueprints
for each country and help to assure that the reorientation of
health services is achieved.

ANALYSIS OF HEALTH SERVICES SYSTEMS - A GENERAL APPROACH

Milton I. Roemer

Professor of Health Services

University of California, Los Angeles, U.S.A.

In every country there is a system of health services, just
as there are systems of education, of agriculture, transportation,
and many other social activities. The health service system is
devoted primarily to protecting and improving the health of the
population by the provision of a great variety of preventive
and therapeutic services; it also has many secondary purposes, such
as providing employment or generating profits, which we cannot
discuss here.

The Scope of Health Service Systems

To define health service systems in this way - as a social
activity for the provision of health services - is not at all to
imply that health services are the only or even the major determinant
of an individual's or a population's health. Socially oriented
physicians and others have recognized for centuries that the
health of people is influenced by the food they eat, the work
they do, the knowledge they acquire, and much more. Nutrition,
occupation, education and other physical and social factors,
furthermore, are mediated in their influence on health by the
biological traits of individuals. Hence a model on the
determinants of health would look something like Figure 1.
I have purposely not tried to show the magnitude of the various
influences, because this is really not known in any generalized
sense. Modern epidemiologists are struggling with this large problem
mainly on a disease-by-disease basis.

Our focus in this conference, then, is on the block in Figure 1
identified as the 'health service system'. As we shall see,
whatever may be its actual influence on the health of people,

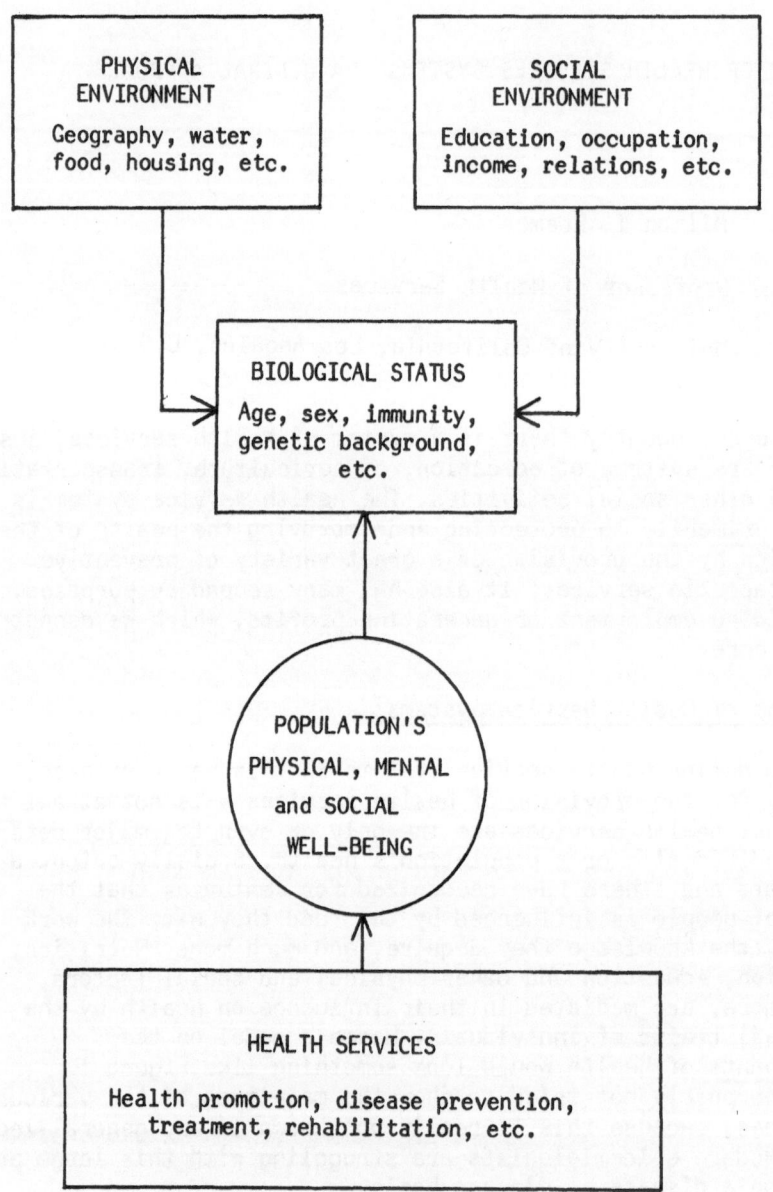

Figure 1. Determinants of Health

men and women everywhere and throughout human history have behaved
as though they believed that health services were an important
determinant of health. In this belief, every known society has
taken actions to cope with disease or injury, to regain health,
and in more recent times to prevent disease and promote health.

The precise ways that this has been done have differed
enormously. In the current era of nation states, the complex of
healthseeking activities in each nation has resulted in a
national health service system. With the different economic,
political, cultural, geographic, and social settings of nations,
their health service systems are naturally very diverse. They
vary enormously in the details of their structure and function,
and hence in their overall complexity. Whatever their degree of
complexity, and whatever may be their efficiency, coherence,
or sense of purpose in providing health services, the
activities in a nation devoted primarily to achieving health
may be defined as a national health service system.

Evolution of Health Service Systems

The current contours of health service systems have inevitably
been influenced by the major historical developments of science
and of society. The steady extension of mankind's knowledge and
control over nature has obviously been a major force determining
not only medical science and clinical medicine but also many
aspects of health service systems; consider the importance of
bacteriology to the development of preventive public health
programmes or the importance of organ pathology and surgery to
the development of hospitals. Outside of the natural sciences,
the evolution of economic orders or methods of exchange of
goods and services have had as great or greater impacts on the
shape of health service systems in the world today. I refer
principally to the rise of capitalist and entrepreneurial
economic orders over the last two or three centuries.

With the decline of feudalism and the rise of free trade,
medical care became one of many commodities and services sold in
the market. This process of exchange initially had distinct
benefits in many settings. By providing physicians, apothecaries,
and others with a source of income, it attracted many gifted persons
into these callings and gave them incentives to work diligently.
It also provided health services for many people in the newly
developed cities, people who lacked the protection of being part
of a feudal estate. The growth of science and universities also
led to the great expansion of knowledge and skills in the
practitioners of the 'healing arts'.

As industry grew, however, and with it democratic and parliamentary forms of government took shape, the concept of health service as a civic or public responsibility also developed. This conception led to consequences in the domain of health services quite different from those associated with the process of free trade. Instead of expecting health services to be bought and sold in a marketplace, mechanisms were developed to provide health care to people on the basis of their human needs and in the interests of general community welfare. These trends were implemented through the founding and operation of hospitals for the poor (later for everyone), the rise of the public health movement, the organization of health insurance programmes, and many other strategies for extending health services to general populations.

Over the last century (i.e. since about 1880) these two concepts of health services have developed side by side. The conflict of values, however, between health care as a market commodity and health care as a social service or even a basic human right has become more and more manifest. With the birth of the World Health Organization and the rise of many equivalent movements in almost all countries, the concept of health care as a right has gained ascendency. Accordingly, complete dependence of health care on market transactions in the private sector is now widely regarded as leading to social inequities and serious deficiencies in health service systems. For these reasons most countries have intervened with freedom of the market place by developing various kinds of collective financing and regulated provision of health service. The degree of this intervention has increased generally over time almost everywhere, but the manner and details of its application have obviously varied greatly. These variations have influenced all component parts of health service systems and, in large part, determine the characteristics of the national systems.

Components of National Health Service Systems

What then are the components of health service systems? Simplistically, they are the many activities that lie behind, that support, and arrange for the delivery of health services to people. But exactly what is a health service? Bed care in a hospital is clearly a health service, but what about care of an elderly person in a custodial institution? Vitamin therapy of a child with rickets is surely a health service, but what about a subsidized lunch programme for all school children? The custodial institution for the aged and the school lunch programme obviously influence the health of people but, unlike the hospital care, this is not their primary purpose.

A health service, therefore, can be best defined as an
activity whose primary objective is health - its maintenance,
its improvement or, if lost, its recovery.

Even with this restricted definition of a health service,
health service systems are complex affairs, requiring many
relationships among their component parts. In a very simplified
form, these parts consist of (1) development of resources,
(2) organization of programmes, (3) economic support, (4)
management, and (5) delivery of health services. The principal
relationships among these components are shown in Figure 2.
If one proceeded no further than this Figure 2 model, I believe
that it would still define correctly the health service system
of every country in the world. Within each of the component
blocks, however, there are many structures and processes or, if
you prefer, subsystems and sub-sub-systems which define the
realities in each nation. The highlights of each of the five
system components may be briefly considered.

Development of Resources. Essential for the provision of
health services are numerous types of human and physical resources.
In their simplest form these consist of: (a) manpower,
(b) facilities, (c) commodities, and (d) knowledge. The
production or development of these resources requires inputs
from various other social systems: education, construction,
manufacturing, and so on - which we cannot explore here. It may
be noted that financing or money is not regarded as a resource;
it is rather a medium of exchange convertible into resources and
services, as noted below.

Health manpower includes physicians, healers, and a variety
of other personnel in all countries. Their formal or informal
training, their precise functions, their work settings, their
inter-relationships, their regulation differ widely among health
service systems. Their number and ratio's to population,
their geographic distribution, and their qualitative level of
performance influence substantially the effectiveness of the health
service system in a country.

Health facilities are also of many types. Historically oldest
are hospitals for the diagnosis and treatment of the seriously
sick. Aside from general hospitals for most acute illnesses and
injuries, there are special hospitals for more long-term mental
illness, cancer, tuberculosis, etc. and for childbirths or the
care of children. Of increasing importance in recent decades are
facilities for the organized provision of ambulatory care - health
centres, polyclinics, health posts, etc. Pharmacies, laboratories,
and even private medical quarters must be counted among a system's
health facilities.

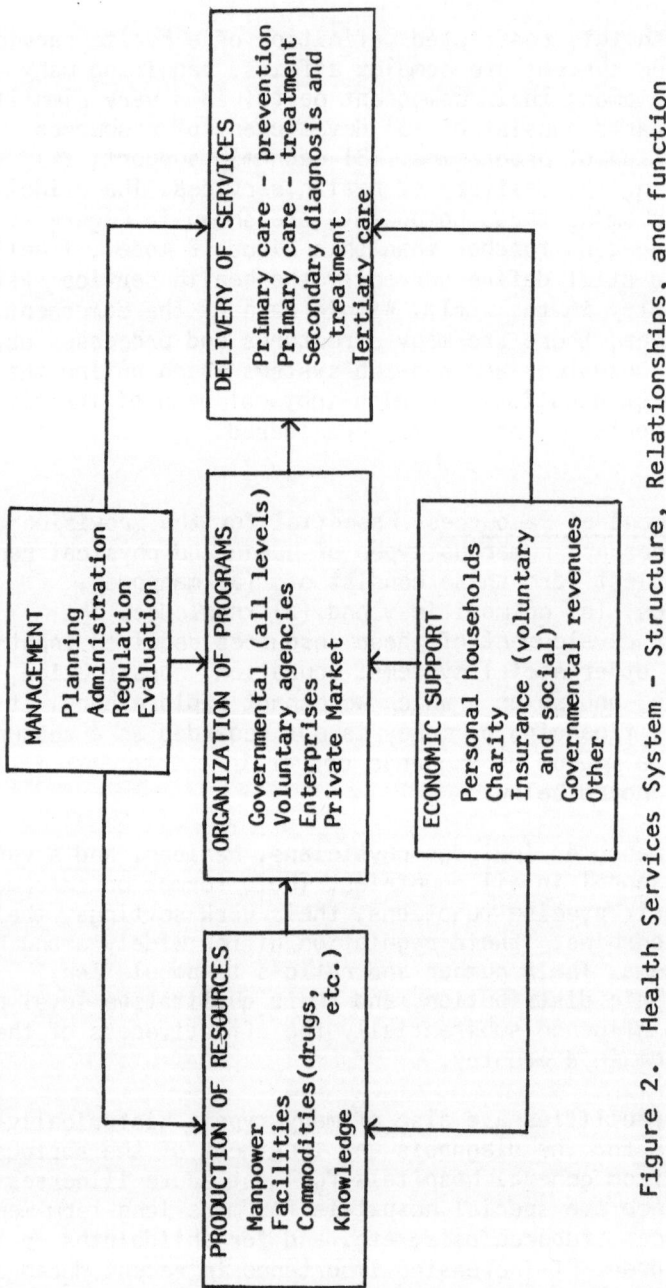

Figure 2. Health Services System - Structure, Relationships, and function

Health commodities are a third type of resource of mounting importance with advances in scientific technology. Best known are drugs and other therapeutic or even preventive substances - behind which stand world-wide industrial establishments. Over the centuries the types of drugs found in nature or synthetically produced have multiplied enormously, and have become matters of international trade involving virtually all countries. Their production, distribution and consumption involve all five components of health service systems. Similar dynamics apply to other commodities, such as diagnostic and therapeutic equipment, prosthetic devices, eyeglasses, wheelchairs, laboratory reagents, bandages, and much more.

Knowledge may not be conventionally regarded as a resource, but it is obviously basic to the operation of every health service system. Scientific knowledge is produced both by observation and experimental research, behind which are further human and technological requirements. Knowledge, like the other three types of resources must, of course, be applied in various ways to result in health services.

Organization of Programmes. The organization or arrangement of resources into various functional relationships or programmes towards certain ends constitutes the second main component of health service systems. The sponsorship of programmes may be governmental, voluntary non-profit, or entrepreneurial, and the proportions among these types are crucial determinants of the general nature of a health service system.

In the realm of government, health service programmes come under numerous agencies, best known of which is the ministry of health. Organized preventive health services are always a responsibility of the ministry, plus an increasing scope of other functions varying among countries. In most countries other governmental agencies are responsible for social insurance programmes financing medical care. Numerous other ministries may have functions for other aspects of the health service system, such as education of personnel, construction of health facilities, environmental controls, etc. Depending on the size (both population and geography) of a country, each ministry may have organized peripheral units at regional, provincial, and local levels, and these units may operate with varying degrees of autonomy.

Non-governmental and non-profit (often charitable) organizations in most countries develop and operate health programmes tackling certain diseases (such as tuberculosis or cancer) or serving certain types of population (such as children or the aged). Voluntary agencies may also operate health insurance programmes. Associations of health professional personnel often represent

their members in negotiations with government and they monitor
ethical behaviour.

The entire private establishment for providing health services
must be considered part of this component of health service systems.
While not 'organized' in the usual sense, it functions through a
market in which the services are bought and sold. The sellers include
physicians, hospitals, pharmacists, etc. and they relate to
buyers (patients) with varying degrees of competition or cooperation.
Health services provided to workers by private industrial
enterprises must also be considered in this private sector.

Economic support. Supporting all the development of health
resources, and their organization into programmes, every health
service system must have various sources of financing. I use the
plural because in every country there is more than one source,
and the proportions of money derived from each source determine
many characteristics of the entire system.

In every country, private individuals or families are a source
of economic support for health services, typically for treatment
of personal disease. Donations to charity are another source,
and this may take the form of donated labour as well as money.
Non-governmental or voluntary health insurance is another source
of great importance in certain countries.

Under government, general taxation is a source of support
for resource development and health services in every nation. The
exact types of taxes (on land, income, purchases, etc.) and
the political levels at which they are collected vary widely, but
everywhere tax funds are used for general prevention and for
medical treatment of at least part of the population. Mandatory
or social insurance is a special form of governmental taxation,
in which the funds are earmarked for a specified purpose - such
as medical care - and they are used only for the benefit of the
persons contributing to the insurance fund and usually their
families. In many developing countries, foreign aid or overseas
charity may also furnish economic support to the health service
system.

The mix or proportions of these several sources of economic
support leads to policies which influence the nature of a health
services system more decisively than any other system component.
It is obvious that support derived from private individuals
channels resources and services to those who have the money to spend,
while support derived form general revenues can be used for services
to others. The dynamics of economic support are, of course, very
complex but they obviously have great implications for health care
equity. On a world scale the proportion (not the amount) of
health-related funds derived from private sources has been

declining and the proportion from public sources has been
increasing.

Management. A second form of support for the operations of
a health service system is management, which includes various
types of social control - including planning, administration,
regulation, and evaluation. Each of the processes may be carried
out with various degrees of informality or rigor. Likewise all
four of the elements of management may be in either the public
or private sector.

Planning may be done at central or local levels of health
service system or in various combinations of these levels. Its
scope may also vary with the types of health activity affected,
vis-à-vis free market operations. Administration includes
several activities in system operations, such as the exercise
of authority, delegation of responsibility, communications,
coordination, and so on. Regulation is usually governmental
but not always; it includes various legal and non-legal forms
of surveillance intended to assure that system activities are in
accord with certain standards. Most regulation has been established
in response to abuses identified in the free market of health
services, and to a lesser extent in the public arena.

Evaluation is a difficult process in any health service system,
because it depends on a flow of information which may be
difficult to arrange. This information should concern the
development of resources, the organization of programmes, and the
delivery of services. Arriving at sound judgements about the
success or effectiveness of all these activities usually
requires statistical data, which may be examined in relation to
standards or objectives. In the absence of such data, evaluation
may be based simply on general impressions of certain informed
observers. By either method, evaluation provides feedback to the
administration in a health service system, pointing to possible
need for organizational changes.

Delivery of Services. Operation of the four components
described in a health service system leads to the final
component: delivery of services to people. These include all forms
of health service: preventive, therapeutic, and rehabilitative.
In terms of the complexity of the delivery process, the services
are primary, secondary, and tertiary.

The types of personnel, facilities, and work setting for
delivery of these services differ substantially among health
service systems, particularly as between industrialized and
developing countries. The differences are also great between the
less organized and more organized types of system - the latter
having health personnel in much more deliberately organized teams

for both hospital and ambulatory care. Within any one type
of system, there may be several different patterns of health
care delivery, particularly for selected population groups or
diseases. Health services for military establishments, for
example, are delivered through highly organized arrangements
in all countries. On the other hand, aged and chronically ill
patients may be served through diverse patterns in different
countries, and also within the health service system of one
country; similar diversity would apply to patterns of care for
patients with mental illness, venereal disease, or tuberculosis.
One pattern of delivery, however, is bound to predominate within
each system type. The extent of deliberate relationships between
and among primary, secondary and tertiary care, or regionalization,
will also vary in different systems.

Determinants and Types of Health Service Systems

The combined characteristics of all five of the health
service system components, just described, define the type of
system found in each coutnry. In the approximately 160 countries of
the world, no two systems are exactly alike, but one can understand
them better by clustering the systems into certain major types.
To do this, one must consider the basic determinants of the
character of national health service systems.

My own observations suggest that the major influences are
economic and political. In addition, there are always several
other influences, which we may group under the heading of cultural.

Economic levels of countries can be quite readily scaled in
terms of their gross national products (GNP) per capita.
Although this index tells us nothing about the distribution of
income in the country, it does describe national wealth.
Countries with relatively high GNPs per capita are, of course,
mainly industrialized, and those with low per capita GNPs are
mainly agricultural. Deviations from those tendencies are seen
in several petroleum-exporting countries, which have relatively
high GNPs without being industrialized, at least currently.

The political characteristics of a country are not so easily
identified and scaled. Although I have been searching United Nations
and other international statistics for a clear indicator of
political ideology - degree of centralized organization and control
and/or degree of governmental (versus private) control of social
affairs - I have not yet found one available for many countries.
In its absence, therefore, I am proposing to scale national health
services systems along a continuum describing the degree of organization of the systems themselves. With these two dimensions,
each scaled into just three levels, we draw a matrix of health
service systems, as shown in Figure 3.

Health Care Policies		
Economic Level	Pluralistic (laissez faire)	Cooperative (welfare states)	Socialist (centrally planned)
Affluent (industrialized)	1. United States Australia	2. Norway Great Britain	3. Soviet Union Czechoslovakia
Moderate (developing)	4. Thailand Philippines	5. Peru Malaysia	6. Cuba North Korea
Poor (under-developed)	7. Ghana Nepal	8. Tanzania Sri Lanka	9. China Mozambique

Figure 3. Health Service Systems, Classified by National Economic Level and Health Care Policies

The cultural influences could be added to the analysis only by adding a third dimension to the matrix, or even a fourth and fifth dimension, which could be shown by multiple matrices. Under the cultural umbrella, one must consider a country's general technological development, which has obviously played a strong part historically. There are many impacts also from religion, community structure, family customs, and language. With sufficiently detailed consideration of all these influences, the world's 160 nations would probably yield 160 different types of health service systems.

With the relatively simple classification shown in Figure 3, the nine conceptual cells are probably sufficiently refined to provide useful understanding of how the main types of health service system work. In each of the nine cells, the names of two countries have been inserted as probable examples. In some of the cells, the systems of many countries would fall, and in others very few. My impression is that the most "populated" cells today are probably 2, 5 and 8. With respect to economic levels, the countries in the top row (cells 1, 2 and 3) have annual GNPs per capita of $ 3.000 or more; in the middle row this figure is between $400 and $3.000 (with the greatest clustering around $1.500); in the bottom row the annual GNPs per capita are under $400 (and principally under $300).

Time does not permit even a brief synopsis of the type of health service system in each of the cells of Figure 3, but a few remarks may be made about characteristics of the systems in each of the three vertical columns. In the first column (cells 1, 4 and 7), the private sector for delivery of ambulatory health care is quite strong. Expenditures for health service come predominantly from private families, rather than government. The resources (all four types), of course, are much lower in cell 4 than in cell 1, and lowest in cell 7. Most health care is delivered by individual practitioners (physicians, health auxiliaries, and traditional healers), rather than teams of personnel.

In the second column (cells 2, 5 and 8), a major proportion of the costs of health service has been collectivized through governmental mechanisms including social insurance. The great majority of hospital resources are in governmental facilities, and in most of these the doctors, along with other personnel, work on full-time salaries. In cell 5, the proportions of populations protected by social insurance are much smaller than in cell 2, but they are expanding. Central governments play a large role in health programme management, but a substantial role is played also by local goverments and local communities.

In the third column (cells 3, 6 and 9), the health service systems are almost entirely governmental. Virtually all resources are within the government, and health services are theoretically available to all residents without cost (except for small

charges made in cell 9). In cell 3, the central government
exercises controls over all aspects of the system, with
somewhat more local participation in cells 6 and 9.
Preventive services are emphasized in all these systems, and their
delivery is integrated with the treatment services.

Even these few highlights of the nine types of health
service system are over-simplifications, but they may suggest
some aspects for differentiation. If any action is to be
taken to improve the health services of a country, in their
quantity, their quality, or their equity, it is obvious that
the strategy would have to differ in all nine types of system.
One would have to take account, moreover, of not only the
economic level and the degree of organization, but also of the
religion, community structure, and other cultural features.

Trends

In a very general way, with some notable exceptions,
national health service systems are evolving towards greater
degrees of organization - that is, from the left to the right
hand side of the Figure 3 matrix. This is seen in trends with
respect to collectivized economic support, health care
delivery patterns, policies of management, and in fact all
components of health systems. To a lesser degree, there are
also trends form the bottom to the top row of the matrix, as many
(but certainly not all) countries undergo economic development.
In health service systems, this is reflected most concretely
by steady enlargement of health resources - ratio's to
population of health personnel, hospital beds, and various
types of health centre. The impacts of these trends are
clearly reflected by world-wide improvements in life expectancy,
infant mortality, and other indices of health. In the years
between 1955 en 1975 the life expectancy at birth in the
developed regions of the world increased from 64.3 to 70.3 years.
Even in the much poorer developing regions, it increased from
42.5 to 53.2. The rates of these improvements are not the same,
of course, in all countries, but the overall trends are still in
a favourable direction. Changes in physical and social environments,
of course, contribute importantly to these trends, but it is also
clear that health services play a substantial part.

Returning to my opening remarks and Figure 1, the health
status of populations depends on many influences beyond health
services. Improvements in health, therefore, require progress
in many sectors of society. One need not wait, however, for
achievement of the enormous changes in a society, required for
greatly improved housing, employment, education, etc., to expect
significant gains in health. More effective and equitable health
service systems can achieve better 'health for all', as WHO puts
it, in a relatively short time.

"A SYSTEMS APPROACH " AND "HEALTH SERVICE SYSTEMS":

TIME TO RE-THINK?

Peter B. Checkland

University of Lancaster

United Kingdom

Introduction

The stucture of the meeting derived from a systems model of a Health Service System. Sessions were organized which covered the major sub-systems of that model and their interactions. This was extremely useful as a device which gave coherence to the discussions, but during the course of the meeting several participants questioned the useful ness of this kind of view, feeling it to be constraining. It is of course one thing to use a systems model to structure a discussion, quite another to assume a real-world Health Service to be the system described in the model. The former action does not necessarily imply the latter.

There was energetic discussion of this point at the meeting. As part of that discussion the paper which follows was written and presented during the Workshop. It discusses some developments of systems thinking during the 1970's which hinge upon using systems ideas as the base of a problem solving process rather than necessarily assuming that the real world contains "systems".

There has been much discussion and debate about whether "a systems approach" has in the past, or could in the future be useful to people engaged in planning, operating, monitoring or controlling what in everyday language can be casually termed "Health Service Systems". Some feel that the systems view of the health area had been very useful, others that the concept (largely assumed to be synonymous with "holism" or a "holistic" approach) has been minimally helpful, providing semantics rather than sub-stantive help; for more hostile critics the systems view is

actually harmful, in that it embodies particular prejudices while
seeming to give them a neutral or scientific air.

This note is a response to that discussion. It summarizes
some of what has been said, and in particular briefly refers to
recent developments in the application of systems thinking to
the messy ill-structured problems of the real-world, a development which sharpens the concept of systems analysis as it
originally emerged by departing radically from it.

The First Systems Analysis: the 50's and 60's

Systems analysis as a service to decision makers emerged
during the second half of the 1950's and can conveniently be
examined in the work of the RAND Corporation. Building on work
in wartime operational research ("operations research" in the
USA) the RAND analysts brought together concepts from engineering
and from economics and developed broad methods of rational appraisal
as a help to decision makers. Methodologically, the approach
started from a definition of the objectives to be achieved, and
defined measures of performance so that alternative means of
reaching the objective or meeting the need could be appraised,
relating costs and hoped-for benefits. This was a highly rational
input to decision making and made no serious attempt to include,
for example, the politics of the situation or the bias of the
decision maker. The decision maker could introduce those in
making his actual choice!

Simultaneously emerged "systems engineering", most articu-
lately from Bell Labs in the USA in which modelling and simulation
led to an examination of the most efficient means to reach an end
defined as desirable at the start of the study. These approaches
were quickly applied in fields other than those of contract-
letting and engineering. Easton's "Systems Analysis of Political
Life", for example, uses essentially these concepts to describe
the nature of political systems. It is characteristic of the
first version of systems analysis that it is taken as given that
there is a need to be filled and that real world arrangements
to do so can be taken to be systems. "System" is given an
ontological existence outside the analysts, and in everyday language
the real world arrangements are described as systems; it is as
if the world were, unproblematically, systemic. The great advantage
of this approach is that it requires a broadening of view from
the analyst, prevents him from being too blinkered, and directs
his attention to connections and interactions.

The New Systems Analysis: "Soft Systems Methodology" in the 70's and 80's

Attempts to apply the "hard" systems thinking of RAND System

Analysis and Systems Engineering in real-world problem situations
during the 70's fell down because the required initial definition
of objectives could not be achieved.

In many real-world problems (such as in providing health
services) there is frequently disagreement over objectives or "the
need to be fulfilled", which is a great part of making the situation into a perceived "problem".

Developments in systems thinking in response to this failing
may be briefly expressed as follows: it was realized that the
systems of concern, so-called "human activity systems" are fundamentally different from physical systems (such as frogs or bicycles)
or from designed abstract systems (such as mathematics). Publicly
testable descriptions of physical and designed systems are possible
such that all observers acting in good faith would have to agree:
"this is so". In the case of purposeful real-world activity,
however, fundamentally different descriptions may persist: one man's
"terrorism" is another man's "freedom fighting", and both descriptions are valid according to models of the world in the minds of
the observers, their Weltanschauungen. In building models of human
activity systems (for which formal methods have been developed)
it is necessary to make explicit the Weltanschauung which makes
that model meaningful. (It should be noted that such is the complexity of human beings that they do not usually hold to pure,
consistent, world images, based on a single pure world view. This
means that models of human activity systems are not models of the
world, they are "ideal types" in Max Weber's sense. Weber's
famous model of bureaucracy is not intended as an account of real-
world bureaucracy - he insists on that point - it is a construct
intended to be useful in examining and understanding real mani-
festations of "bureaucracy" in the everyday world. Models of human
activity systems are precisely ideal types for use in setting up
a rich debate.).

These characteristics of human activity systems dictate the
shape of "soft systems methodology" (2). It starts from a problem
situation (never a taken-as given problem) and names a number
of holders of roles in the situation as "problem owners" - for
example in an urban renewal program, the problem owners could be
taken to be the planners of such a program, the professionals
who will be carrying it out, the society as a whole which funds
it, or the people living in the decaying area who will be the
program's beneficiaries or victims. In the light of these definitions of problem owner, some human activity systems hopefully
relevant to the problem situation are selected (avoiding thinking
in terms of "solutions to problems".) It is a good idea to select
relevant systems which are manifest in the real world as organization
structures, and especially important to select some systems which
are not. For example, in a hospital entailing both administrators

and clinicians, it would be useful to take as relevant system
"a system to ensure communication between the two groups" although
this will not map on to any part of a formal organization structure.

Once the relevant human activity systems have been selected,
they are carefully named (in so-called "Root Definitions", and
from these Definitions models of the systems named are built
(so-called "Conceptual Models"). Techniques for doing this naming
and model building have been developed.

The models are then used in the problem situation to
structure a debate about changes conceivable in the situation
under study. This is done formally by comparing the conceptual
models with real-world happenings. The aim of this debate (which
often leads to new, richer choices of "problem owners" or "relevant systems") is to find some changes which meet two criteria:
that they are systemicly desirable (given this particular choice
of relevant systems) and that they are culturally feasible for
these particular people in this particular historical and political
situation. The debate itself often orchestrates both the consensus
and the conflicts in the problem situation.

Given the definition of the desirable/feasible changes the
"new problem" is now to implement them in the problem situation -
the cyclic learning process can begin again as a new situation is
created (figure 1).

Note that the model building is not normative or utopian; the
models are not necessarily intended as designs. Thus, a priest
making a systems study of the Roman Catholic Church ought to use
as one of his Root Definitions Marx's remark about religion being
"the opium of the people". He may not agree with it but its use
in the debate will help the generation of radical new thinking.

The methodology as a whole is a cyclic learning system aimed
at generating insights which lead to purposeful action. It is
fundamentally different from RAND systems analysis or systems
engineering, a point to be discussed in the final section.

Two Paradigms of a Systems Approach (1)

The fundamental difference between soft systems methodology
and the earlier Systems Analysis lies in their attributions of
systemicity. To RAND analysts, to systems engineers and probably
to most people with a casual or non-professional interest in a
systems approach at the present time, it is taken as axiomatic
that the world is systemic, containing systems of various kinds.
Natural systems can be investigated and understood; designed
systems can be engineered. The approach to understanding or creating
these systems, it is assumed, can be systematic - that is to say
rational, well-ordered.

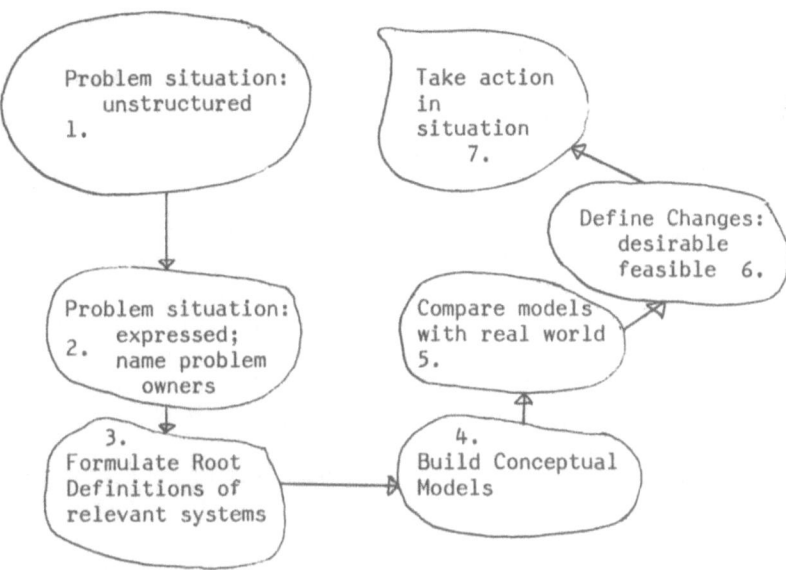

Figure 1: Soft Systems Methodology as a cyclic learning system
(Ref. 2)

R: reality
M: the approach
   to understanding
   reality

Paradigm 1: R is systemic
            M can be systemic

Paradigm 2: R is ?
            M can be systemic

Figure 2: Two Paradigms of s Systems Approach (Ref. 1)

To soft systems methodologists, however, it is not taken as given that the world contains systems. The world is taken to be highly complex and ultimately mysterious: but the approach to trying to understand it, it is assumed, can be systemic, using intellectual constructions of a systems kind (the conceptual models of stage 4 in figure 1) to compare with reality in order to gain insights as part of a cyclic learning process which in itself is systemic: it is a process which learns.

Figure 2 illustrates the two paradigms. At the practical level they are embodied in "systems analysis" and in "soft systems methodology". At the level of their underlying philosophies - which they take as given - Paradigm 1 is an example of positivism, Paradigm 2 is an example of phenomenology, of the hermeneutic circle. The former yields primacy to the given world as known through experimental evidence; the latter yields primacy to the mental processes of observers rather than to the external world. The former is an ontology, the latter an epistemology.

The two paradigms of a systems approach are complementary to each other, but the newer paradigm is likely to be the more useful in any problem situation dominated by the meanings which actors in the situation attribute to what they observe. Health service systems can usefully be characterized by the older paradigm, in order to describe them in a consistent and complete way. But problem solving within them, as in other ill-structured problem situations, is likely to require the systemic learning cycle of the newer approach.

References

1. P.B. Checkland, "Rethinking a Systems Approach", Journal of Applied Systems Analysis, 8, 1981.

2. P.B. Checkland, "Systems Thinking, Systems Practice", John Wiley, 1981.

II. TOOLS FOR DOING IT

Albert van der Werff

Ministry of Health and Environmental
Protection, Leidschendam

National University of Limburg
Maastricht, The Netherlands

Introduction

The reorientation of health services systems needs
capable management which is able to understand the pressures
within systems and the relations between systems and the
changing environment.

All country analyses (which will follow in section 3)
confirm that the national health services systems are entering
an era now with a lower rate of economic growth than in the
previous postwar period. This new economic situation will
certainly have deeply felt consequences for the development
and reorientation of health services systems. Alongside
this there has been growing dissatisfaction with the
lowering effectiveness of health services systems. As a
consequence health policy making has become very much in
the public eye. In this connection also the role of government is changing in many countries.
Thus, the following changes may be expected.

Available resources will no longer grow as they used
to. New goals may have to be formulated for the deployment
of these resources. Expressed in terms of health functions
this will mean in almost all countries that the present
"preventive-curative care mix" with its emphasis on curative
care should be reoriented in favour of health promotion and
disease prevention. And the existing health services
complexes should be designed and consequently be redesigned
and reoriented as "systems supportive to primary health
care".

Changes of this kind certainly require the application
of a systems approach. This argument is undoubtedly strong.
On the other hand under circumstances of severe economic
pressure governments (administrators or managers) may look
for short term savings neglecting the effects of their
decisions on other components within the system or
the consequences for the longer term. So there may be
a conflict situation with respect to the right balance
within systems at a certain point in time and between
short-range and long-range aims. Changes to be effected
at one point in the system may have consequences elsewhere, which are not anticipated.
One may expect that these conflict situations may provide
severe obstacles to efforts to reorient health services
systems. One may also expect that the parties concerned
will react more defensive as they become more afraid
to loose positions. These problems of low effectiveness
of health care in a limited-growth economy and a defensively
reacting society cannot be solved without major changes
in the managerial processes and mechanisms by which health
services systems have to be reoriented.

Below various tools for systems management will be
discussed by different experts in the field. Successively
the following issues will be dealt with: planning, manpower,
development, financing, cost-effectiveness, information
services and research and development.

SOME ASPECTS OF ALTERNATIVE APPROACHES TO PLANNING

A.W.M. Meijer

State University of Limburg

The Netherlands

0. Summary

This paper, written by a Dutch sociologist, is a venture in adding some apects of the formal theory of planning to the descriptions in this book of the characteristics of the health services systems in many countries.

The study provides a short overview of planning theory. First, the need and role of planning in modern society is mentioned, followed by an enumeration of planning dimensions (par. one). The aims and objects of planning are described briefly. The objects of planning are considered as five planning modes (par. two). The distinction between theory in planning and theory of planning is made, while two other problems of planning theory (human growth and the planning process and planning , decision and control) are discussed (par. three). The next chapter deals with some aspects of the relation between planning and the systems-approach: variety, complexity, uncertainty and rationality (par. four). An overview of styles and types of planning has been given in paragraph five, and attention has been drawn to future-types of planning. These new types will be needed in a situation of economic stagnation with which they must correspondend to be helpful. The paper is concluded by an evaluation and a list of references.

1. Introduction

1.1 The need for planning

It may be a good thing to start a paper on planning with

a short description of the position of planning as a social
activity.

The great many contributions of the recent years on
planning show an increasing interest in planning as a social
activity and a tool for policy making, especially with regard to the increasing complexity of processes of management
and control. Five reasons for planning-activities are mentioned
(in 't Veld, 1980 II):
1. The extension of the control-processes and the increasing
   time lag between decisions and the appearance of effects
   of a certain decision requires planning. This is the time-
   aspect of planning.
2. The position and role of the state in the modern society
   has been changed, which showed the need to develop instruments to control the future. What is pointed out here is
   the control-aspect of planning.
3. Planning is as useful as an instrument for policy as legislation and financing is. Here, the instrumental aspects of
   planning are emphasised.
4. The control-system has become so complex in itself, that
   planning of this system itself is necessary. Here, the
   integrative aspect of planning has been stressed.
5. Planning will contribute to more rationality in the
   activities of people and organisations. This rationality
   can refer to the rationality of ends or the rationality
   of means. This aspect is accentuating the rationality-
   aspect of planning.

Usually most emphasis is given to the aspects of rationality.
But in recent publications more and more attention has been
given to the other arguments of planning.

1.2   The role of planning

As the arguments for planning are different, so are
there differences in the role of planning in each societal
context (in 't Veld, 1980 II).
First of all, planning has implications for social relations.
The next step is that planning has to be considered as a
social process in itself. A third variant is that planning
becomes an institutionalised activity in social relations.
That means that the object of planning will be extended to
the social order itself. The last step may be that planning
is inherent to every social activity and can be made explicit
as such.
It will be clear that planning has no single solid theory
or is no uni-dimensional instrument since there are so many
arguments for (and against) planning and since planning has
or may have so many roles. This point will not be discussed
in this paper.

1.3 Dimensions of planning

     To facilitate discussions on planning, it is necessary
to have an analytical framework.
In paragr. three a distinction between procedural and substantial planning will be made, i.e. the difference between
the theory of planning and the theory in planning. In most
theories of planning (called planning theories) three
dimensions of planning have been applied (Faludi, 1976,
Rade & De Smit, 1977).
a. Blue-print-planning versus process-planning: blue-print-
   planning is a type of planning in which action is undertaken on the base of a complete description of a final
   result at the end; while process-planning is a method of
   planning which is allowing for stepwise adaptation of
   the intended course of action each time when additonal
   information is indicating so (learning by doing);
b. normative planning versus functional planning: normative
   planning is planning by ideals- and values-setting, seen
   as the most important phase in planning; functional planning
   is planning on the basis of given objectives;
c. rational-comprehensive planning versus disjointed-
   incrementalism: rational-comprehensive planning indicates
   a mode of planning based on an ideal view of decision-making
   as a rational activity of making the best choices out of
   all possible objectives and means and on a strong will to
   take all aspects of reality into account (holistic view).
   Disjointed-incrementalism departs from the opposite
   situation.

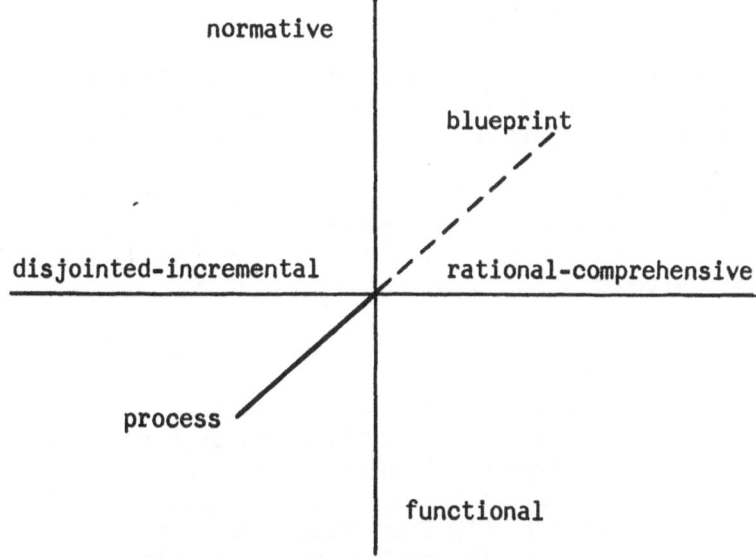

Figure 1: Three dimensions of planning (Faludi 1976,
          Rade & De Smit, 1977)

The first dimension will be discussed in paragraph three; the second in paragraph four and the last in paragraph five. In the next paragraph attention will be given to the aims and objects of planning.

2.   Aims and Objects of Planning

2.1. Aims of planning

In the first paragraph the need for and roles of planning are shortly discussed. The next paragraph will pay attention to the problems of planning theory and the fourth paragraph deals with systems and planning. Earlier was mentioned that the role of planning is still changing. The same applies to the aims and objects of planning in social systems. Thoenes (1980) distinguishes three catagories of aims of planning:

1. planning for development: in our opinion a rational-comprehensive type of planning, based on a great believe in progress and a strong will to reach a certain future (normative and blueprint planning-model):
2. planning for perfection: also a rational type of planning to complete and improve an existing system (functional and blueprint planning);
3. planning to avoid disasters: an incremental type of planning, in my opinion, often with only one special target: to keep the system going. This type of planning deals mainly with symptoms of disfunctions in the systems (functional and process-planning).
To this list two other categories may be added, in our opinion:
4. planning for restructuring: a type of planning which is a mixture of process planning and incremental planning. In this case planning is used as an instrument, usually based on legislation;
5. planning for decline: a rather functional but radical type of incremental planning, to cut parts of the system (or even the whole system) for different reasons (ideologic, financial) without specific selection-procedures.

In the Netherlands, a mixture of all these aims exists in health planning.
Mostly, planning is set up to develop or perfect the health system, even in situations in which the system seems to be complete. Such a type of planning is very common at the regional and local administrative level for political reasons. This is understandable (but often unrealistic) in situations in which the health services are still financed by social and sickness funds. On the central level, it some-

times seems that planning is used as an instrument to
keep the system going, although planning has increasingly
been used for restructuring the health system.
These are only small starts to give the system new clear
normative directions. One may say that rational-comprehensive
health planning never has existed in the Netherlands.

2.2 Objects of planning

Two major characteristics of the process of moderni-
zation in industrialised nations are the process of
generalisation of values and the process of structural
differentiaton. The former process means that the
range of rationalised values in society is getting smaller;
the latter points out that a division of functions and
activities may arise creating a differentiated pattern of
specialized institutions, which has to be coordinated and
integrated to obtain the same effects as before. This process
of extensions and externalisation (Hall, 1977) however, may
make coordination and planning necessary, but complex. The
use of this concept of structural differentiation (and
extension) enables a distinction of five modes of planning,
ranked by their objects:
1. operational-planning: the planning of activities and
   functions (services) themselves (the primary process);
2. structural planning: the planning of means and structures,
   sometimes as networks of the services themselves (not
   the institutions);
3. strategic planning: planning which focusses on objectives:
   the innovation of the structure and operating services
   with emphasis on implementation aspects;
4. institutional planning: planning which deals with competencies:
   the legislative aspects of services, structures, institutions
   and networks;
5. normative planning: the heuristic, more or less utopian type
   of planning, providing new guidance for other types of
   planning by setting ideals and values.

These five types of planning can be considered as a
hierarchy (or as a figure of concentric circles) with the
first type (operational planning) at the bottom of the
hierarchy (or as the central circle).
The first type of planning and the last one are the
oldest. Operational planning may be seen as a form of
functional rationality, mentioned in paragraph one.
Normative planning focusses on the future over a broad
range of concerns with the intent of changing the
behaviour and outcomes of the system (Blum, 1974).
Strategic planning has been applied more recently. It
focusses mainly on the problem of changing systems.

normative planning	why?	ideals and values
institutional planning	who?	competencies
strategic planning	how?	objectives
structural planning	where?	means and structures
operational planning	what?	tasks

Figure 2: Five levels of planning

Attention to institutional planning has been paid since growth of social systems is no longer selfevident. This attention is increasing in particular in those situations in which a lack of money and means arise.

3.   The Problems of Planning Theory

3.1. Substantive versus procedural theory

In recent planning theories a distinction is made between theory in planning and theory of planning (Faludi, 1976, Van Doorn en Van Vught, 1978, Salet, 1980). The former theory, called substantive theory, is concerned with the planning in a specific sector (a.o. health planning, urban planning); the latter, the procedural theory, concerns three major issues:
1. understanding planning, its objectives, agencies and its procedures;
2. comparing their different forms and transferring experiences from one to the other;
3. designing and improving planning agencies and planning procedures. This third issue is often called meta-planning, which means the planning of the planning.
Planners should view procedural theory as forming an envelope to substantive theory (Faludi, 1976).
The distinction between both theories is risky because substantive theories often have their own procedures. Procedural theories may also have their own (implicit) presuppositions. Nevertheless this distinction is made to emphasise that the text of this contribution mainly is concerned with the procedural theory.

3.2. Human growth and the planning process

Planning has always been understood as preparing and
implementing action in an intelligent and rational way.
Planning promotes human growth by the use of rational
procedures of thought and action. This requires prior
discussion on the concept of human growth itself and
the reasons for putting it forward as an ideal.
Planning promotes human growth as a product in different
ways.
Firstly, by identifying ways of attaining ends; secondly,
by contributing to the learning of the future; thirdly,
by regarding growth as a process. In this sense, planning
is not a product but a process and in that way an instrument
for human growth (Faludi, 1976). Since the rational planning
process in itself is a part of the process of growth, it will
be clear that there is a constant and close relation between
both aspects. The recognition of this relationship leads
to two conclusions:
1. more attention has to be given, to the question how
   information is gathered, to the learning process and to
   the development and use of knowledge, and
2. the excessive expectations of planning as an instrument
   to bring us to an ideal and desired future are minimized
   into more realistic ones.
The single-sided orientation on how the future ought to be, has
been changed into an orientation on how the future may be.
On the other hand, sometimes theories in planning keep a
great distance to all normative (or evaluative) aspects
of planning, of knowledge and of human growth.
In such views, planning is mainly considered as a component
in a regulatory process and no distinction may be made
between external and internal goals. The concept of planning
as an instrument for human growth that such, in fact
normative, suppositions are maintained in the end.

3.3 Planning, Decisions and Control

Planning as a rational process of preparation and
implementation of action deals with decision-making and
control. In most cases the planner is not the decision-
maker and even not the person who has the task to implement.
Very often the planning process passes through many hierarchical levels. For that reason a distinction is made in
planning-theory between ends, decision-variables, conditions,
constraints and means. Especially in situations in which
many planning-levels and many different agencies exist,
emphasis is on the decision-making process and the process
of control.

In the Dutch situation for example in health planning
one cannot speak of one decision making body, covering
one object to plan and possessing a solid control
mechanism. The Dutch Government, for instance, as one
of the legal health planning agencies, is not a direct
provider of health services. She is only the indirect
provider of the majority of the financial resources in
the health field. Her major task at this moment is to
legislate the conditions under which the health
system can function in a rather autonomous way.
In such circumstances, it is impossible to speak of one
concept of planning, valid for all situations. Here again,
the close mutual relationship between a substantive and
a procedural theory is visible.

4.   Planning and the Systems-Approach

4.1. The systems-approach in planning

An important tool to planning is the systems-
approach. In the systems-approach realitiy is considered
as a system, consisting of collections of components (or
subsystems) with mutual relationships. Those components
and relations are directed to certain ends. Every system
has an environment. In most cases systems have relations with
that environment, in such a way that the environment influences
the system and/or visa versa. For the description of reality
as a system, a model of that reality has to be developed. The
components and relations in that model may be objects of
planning. The main problem is to describe reality in terms
of components and relations To do this well and before planning
can be initiated some problems have te be faced (Van Doorn &
Van Vught, 1978). These problems are the problems of
variety, complexity, uncertainty and rationality.

4.2. Variety

The description of reality in terms of components
and relations means a reduction of the variety in reality.
In a model only a limited number of components and relations
can take place. The selection of the characteristics of the
system is a problem of evaluation and knowledge which is
mentioned in paragraph three. The variety of a system or
a model can be measured by the following major factors:
1. the number of components (units of subsystems) in
   a system;
2. the relative differentiation in the selected components,
   and,
3. the degree of interdependence between those components.

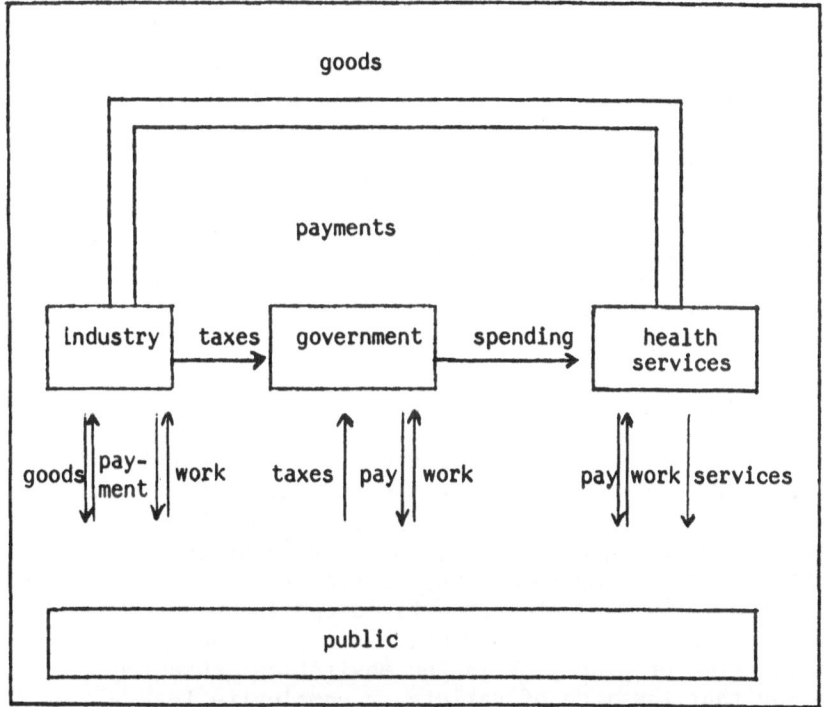

Figure 3: The Health Services and Society, a Model.
In this model, variety is relatively low. There are only four
elements (subsystems), a low number of relations and both
are not very differentiated. The interdependence between
the components is weak.

4.3. Complexity

A system is supposed to have a goal. Usually a
system has more than one goal. Very often a distinction
is made between goals, objectives, decision-variables,
conditions and means. The usefulness of such distinctions
depends on the level of the (sub)system. A decision-
variable on a higher level may be a mean or a condition
on a lower lever. Sometimes a condition may change into a
variable in a certain phase of the planning process.
Complexity in the planning of systems exists in case:
1. the relation between the objectives is of such an
   order that maximalization of the chance to reach one
   leads to minimalization of the chance to reach the
   other(s);
2. constant uncertainty is existing about the validity
   of the objectives as a result of inaccurate or
   incorrect information or of a turbulent environment
   of the systems;
3. the capacity and competency to take decisions on the
   planning object is spread over many individuals or/
   and agencies (Steinbrunner, 1974).

4.4. Uncertainty

Planning has to do with reducing uncertainty. First
of all, there may an uncertainty in the reduction of
variety: do we select the best characteristics of the
system? Secondly, the question of complexity leads to
uncertainty: is there any chance that there are
contradictions in the selected objectives and in the
decision-making structure and process? Third, is the system
changing in time or place, or is the environment changing
in such a way that the mode of variety or complexity is
changing?

4.5. Rationality

In planning theory two kinds of rationality are
distinguished: functional rationality as a form of
thought which is rational with respect to means, not
to ends, and substantial or value rationality, concerned
with basic considerations of human purpose. Weber,
Mannheim and modern sociologists are argueing that
modern society is concerned only with the notion of
functional rationalization. Mannheim has stated that
functional planning remains on the level of functional
rationality and fails to consider alternative ends
(Mannheim, 1940).
Since functional planning is keeping the ends of a

program constant, planning tends to proceed in a
blueprint mode and vice versa. In the same way, a
planning process with multi-planning agencies or levels
tend to be functional planning (or planning in a
functional mode) because each agency or level is keeping
his ends constant and can only decide on and keep
control over a limited number of variables. Such
planning may be characterised as incremental planning,
with a blueprint ahead, that never will be reached
(Faludi, 1976).

On the other side, value or substantional rationality
often tends to normative planning. Normative planning
is chiefly concerned with the ends of action of a
social system. The goals of normative planning are
those of the system itself. In that way normative
planning refers to an integral or holistic approach
of planning. Such a planning may be utopian or
dictatorial. It will be utopian when we do not have the
broad range of information about great and complex
systems, needed for achieving our ends. It may be
dictatorial, if the decison-maker will neglect this
fact and tries to reach his ends against all criticism
and resistance.

In conclusion, neither functional rationality alone,
nor substantive rationality, will lead us to a way of
planning that will satisfy totally.

5.   Types of Planning

5.1. Styles of planning

A considerable number of planning theories and
types of planning have been put forward in the literature
over the recent years. Although there are great differences
in the descriptions of the way planning is done or ought
to be done, four main styles of planning can be distinguished, each existing of one or more planning types.
The first style may be called the orthodox style of
planning, that means a style of planning, focussing on
the ordering of objectives. The three main types of
planning within this style are: the rational-comprehensive
type of planning, the disjointed-incremental type of
planning and the third type: the mixed-scanning type.
The second style is the system-rational planning style
or cybernetic planning-style, which focusses on planning
in terms of functions. Great emphasis is put on aspects
as cohesion and coordination. Less attention is spent on

the process of searching for objectives and on the role of
the persons of institutions in charge.
The third planning style may be labelled <u>communicative</u>
<u>planning</u>. This style of planning focusses on the
consequences of decisions, means and objectives for the
participants. Much emphasis is given to the decision-
making-procedure and -structure. This type of planning
is a functional one, often ending in a normative type of
planning.
The last style, which can be distinguished, is the <u>indicative</u>
<u>or adaptive planning style</u>. In this style, planning is
seen as a process of objectives-means. This planning style
is stressing processes of change.

5.2. Orthodox types of planning

In the last paragraph three types of orthodox
planning were listed. In this paragraph these types
will be discussed.
These types of orthodox-planning are:

A. The rational-comprehensive or synoptic planning model

This approach may be characterised by roughly four
classical elements (Hudson, 1979, Aquina, 1972), which
assume the possibility of
1. goals setting in terms of ranking and clearness;
2. identification of policy alternatives in terms of
   availability and capability;
3. evaluation of means against ends;
4. implementation of decisions.

These four elements together mean that all variables
and all causal relations between those variables are known.
There are no problems of variety, complexity or
uncertainty. Every step is known beforehand and all
actions are directed towards certain aims.
The main objections, frequently found in literature,
against this planning type are:
1. total knowledge is a fiction, especially when more
   complex systems are the object of planning;
2. social life has some dynamics. It is impossible to
   have complete certainty about all aspects of complex
   dynamic systems. Therefore, comprehensive coordination
   is not feasible;
3. there is no place for subsystems or elements in the
   system with a certain kind of autonomy. This makes
   synoptic planning so simple that one can argue that
   there is no need for comprehensive planning. But
   decisions can not always be made by order;

4. there is little or no attention to the context or
   environment of the system;
5. priority is given to methods of thinking in terms
   of causal relations although there are many other
   relations;
6. the planning (or decision-making) sub-system has not
   always the possibility and power to steer and control
   the total system;
7. objectives are often multi-interpretable;
8. objects and means can often be exchanged, depending
   on the planning level.

The main weakness of this type of orthodox planning
seems to be that experiences from the planning of technical
material objects are transferred to the planning of social
systems. (Peper, 1974). After Johan Galtung this can be
called "a fallacy of the wrong level" (cf. par. 2.2.
objects of planning).

B.   The disjointed incremental type of planning

This planning type is the opposite of the rational-
comprehensive type. In this model the process of policy
is seen as a continuing story of multi-interpretable
objectives and values and changing power-structures.
The policy processes are fragmented (disjointed) and
changes caused by policy are marginal.
Policy-makers do have "a limited view and less time"
which makes that of all alternatives only those are
investigated which differ to a limited (incremental)
degree from existing policies.
The objections against this type of planning are the
following:
1. problems are not solved but attacked;
2. decisions are always marginal to the status quo;
3. objectives are adjusted to the means;
4. problems are reformulated again and again until an
   available solution is found;
5. analysis and evaluation is done by all parties in
   an uncoordinated (disjointed) way.

The main result of this planning is that it has a
certain content of reality. Very often policy making
seems something like "muddling through" (Lindblom). But
the model seems to have some pre-suppositions:
1. incremental planning is good planning because this
   is the only way to carry out planning in reality;
2. incremental planning gives possibilities for flexible
   planning;
3. incremental planning brings us always to consensus;
4. incremental planning has nothing to do with power-
   structures.

C.   The mixed-scanning type of planning

       This type of planning, first described by A. Etzioni
(1967),offers an approach in which a broad scan of the
field of interest is undertaken to identify which decisions
can continue to be taken incrementally and which ought
to be taken rationally. Etzioni claimed that by varying
the detail of the scanning activity, such an approach
would be flexible enough to deal with a wide variety
of contexts and environments of varying stability
(Wisemann, 1979). This type of planning deals mostly
with objective-setting, but also with implementation.
In the first phase of the mixed-scanning model, there is
a free exploration of all alternative objectives, but
those which are unattainable because of normative of
political reasons are selected out. The next phase is to
prepare implementation. All activities, which have to be
done are set into steps of actions. To each step the
necessary means and supports are allocated. Those steps
(of action), which will be without means or support, are
selected out.
The last phase is a permanent evaluation of the
implementation of the decisions.
According to Etzioni these methods will give a society
a tool to handle its problems more effectively, because
the model has a higher capacity to build consensus (than
the incremental model), and a lower necessity for
different means of control (than the synoptic model).
The model of mixed-scanning is a real mix of the
models of rational comprehensiveness and disjointed
incrementalism. For the objective-setting all
objectives have to be investigated. There is also an
open selection of means to serve a goal. The choice
of the objectives by the decision maker is based on
"normative integrity, which should be encouraged".
(Mastik, 1981). An important role is given to the
intellectuals to pry over the walls in which society
tends to box itself. Etzioni asks for an active
society in which powerless groups will be mobilised
and can participate in the decision-making process.
This last point leads to the styles of planning in which
less attention is put on the rational aspects of
planning. But before discussing these aspects, some
notions on the orthodox planning will be made. Orthodox
planning is based on three conditions (Van Gunsteren,
1980):adequate knowledge and information, a reliable basis
of power and cooperation, and a stable and obedient field
of implementation. Knowledge is growing but the complex
problems, which have to be solved by planning are also
growing. In addition, the growth of scientific knowledge

has become a process of structural differentation (as
discussed). Therefore, the problem of integrating
knowledge is arising and this problem is ever growing.
And above all, reality is changing every day, so that
knowledge of today may be nonsense tomorrow. Orthodox
planners, who trust on knowledge absolutely, may fail
for this reason.
Orthodox planners want rationality. But political
and social life in many countries is based on all
kinds of networks of implicit balances. Explicit
rationality of any kind will disturb these networks.
Planners will come in conflict with politicians.
Orthodox planners are also asking for stability and
flexibility for implementing their plans. Modern
organizations,however, are not stable and flexible.
So the plan will fail but the decision-maker is no longer
interested since he made the decisions beforehand.
As these three conditions usually are not met, orthodox
planning usually will fail. Orthodox planning is nothing
more or less than changing chaotic failures into organized,
systematic failures (Van Gunsteren, 1980).

5.3. System-rational planning

Some planners state that orthodox planning is not
systematic enough. They try to look not only to subsystems
but also to systems of subsystems. Such systems are
organized, complex wholes, built of subsystems. A social
system has a structure (components and relations) and
a culture (values, norms and expectations). Attention
is given to those subsystems, which regulate the
conditions to which (sub-)systems can function, but the
way by which these conditions are formulated are kept
out of sight in the analyses. Much attention is given to
the dynamics of systems and to the environment of (sub-)
systems. A clear division is made between normative,
strategic and operational planning. The structural
and institutional aspects are mostly given as the
results of the former planning-modes. "Leit-motive" for
system-rational planners are coordination and harmonization.
For to coordinate and to harmonize, often many forms of
institutional autonomy have to be broken down, wich
is a critical point in the implementation of this type of
planning. Another critical point is the description of
the dynamics of a (sub)-system and the expected (or
desired) direction of these dynamics. These are normative
points to which system-rational planners do not have
much influence. In the cybernetic variant of the system-
rational planning much emphasis is put on the control
of the variety of the system. Control is nothing else

than to blockade any interference or disturbance of
the variety of a system (Kickert, 1980).
Both variants of the system-rational planning types
do assume a sub-system, which is able to control the
system. The objectives of this control system are
mostly out of sight, except the objective of (self-)
regulating of the system. This type of planning
also does not give any answer to the normative
questions about ends and means in so far as these
questions are not linked to the continuation of the
(sub-)system.

5.4. Communicative planning

Another type of planning is the style of planning
which puts attention to the aspects of participation
and communication in decision-making processes.
Communicative planners do not have the opinion that in
modern life enough control and coordination structures
and subsystems exist. Communicative planners are
looking for political-rational decisions. Participation,
in this case, is a value in itself. Democratisation is
a form of a powerless communication in decision
making processes. Until now, the experiences in the
Netherlands with this type of planning are not very
encouraging (Van Gunsteren, 1980). A reason for these
failures is that the democratisation and communication
of complex topics do assume a certain level of knowledge.
Very often this level is not reached by a great number
of participants. Their reactions are therefore not
based on substantial rationality but on emotions or
values (depending on the issues). On the other hand,
the type of planning concerned has often been tried
out in situations in which democracy is absent (or
even a fake).

5.5. Indicative planning

In France, after World War II, there was a certain
type of planning that emphasised the flexibility
and the stability of the object of planning. On the
basis of a voluntary cooperation of all important
parties involved in the object, a rough plan with
regulations was made. This type of planning, called
indicative planning, starts from the principle that
planning must be a learning process because
the future is always uncertain to some extent. The
main question in planning is: how can we be prepared

for the future, not only at the top of the operating
system but at all levels involved.
Indicative planning is planning by global and rolling
plans, mostly compared with regulations or guidelines
from the top to the bottom. The central idea in this
type of planning is the idea of collective consultations
of the parties involved on the objectives, which have
to be reached within a certain period. The technical
preparations for those consultations are made by an
independent planning-office, but the general objectives
for the plans are determined beforehand by the administration or/and parliament, often after consulting
some advisory bodies. The elaboration and implementation
of the plans is commissioned to the private sector (the
services themselves) per sector, per region or in
covering committees. Indicative planning therefore may
be considered in a sense as a mixture of the mixed-
scanning planning type and communicative planning.

Against this type of planning some objections are made
(Wentink, 1976; Van Gunsteren, 1980).
Firstly, this type of planning gives much power to the
bureaucratic planning offices. The experts of those
offices may have a solid grip on the procedure and on
the content of the planning. Sometimes this grip is so
tight that it becomes impossible for politicians to give
the planning process another direction against the ideas
of the bureaucrats.
Secondly, one of the major goals of planning is to create
certainty. An ever-rolling plan does not bring much
certainty to the participants, even if it is a global
plan.
Thirdly, this planning type seemed to be rather successful
in times of (economic) growth, but it may not work
so effectively in times of (economic) stagnation or
decline.
Fourthly, a problem is to bring the learning processes
of several levels into one decision, based on consensus.
Also the feed-back to those levels in the form of
regulations and/or guidelines is complicated. These feed-
back forms will not always be accepted and implemented,
but lead to new propositions for change (compromise).
The experience with indicative planning in the health
field in the Netherlands (province of Noord-Brabant) is
not very encouraging but certainly not hopeless. A problem
of importance for participants seems to be the level of
abstraction of the global plan and the duration of the
planning period between the start of the process and
the first implementation. In fact, these are "classical"
problems of planning.

5.6 Future types

In some welfare states, some new types (or styles?)
come into sight. Recently corporate planning, strategic
planning, scenario-methods and other kinds of "business-
activities" have been implemented in the public sector
(Van Gunsteren, 1980). The stagnation of economic growth
and the lack of finance have been quoted as the main
reasons for these developments. One may doubt, however,
whether these technologies can be developed and implemented
successfully in the health sector. Today the values in
society related to health (growth) are certainly not yet
in line with the changing socio-economic environment
(stagnation). So it may be that the first tasks for planning
(and planners) are to find methods to develop those
changes of values. It is clear that new planning types
only will succeed if they fit the societal context.

6.   Evaluation

Planning has become an important factor in the public
sector. A great number of problems in the welfare state
could and can not be solved without planning. However,
planning has not been very successful. One may conclude
that planning did not yield the results which were
expected: neither results in the sense that planning has
solved most concrete problems, nor results in the sense
that planning produces always consensus in a democratic
way. It is possible that the wrong modes of planning have
been applied to certain problems.

Since 1945 there have been changes in styles and
types of planning, especially in the social planning and
in town-and-country planning. In paragraph 5 an impression
of those changes has been given.
Research on such changes in health planning are less
numerous. A reason for this lack seems to be that
health planning in many Western countries still
shows a relatively low level of development. In those
countries where some planning activities on health
care have been started, there is a general tendency
to broaden the scope of planning from institutions to
health sectors and from health sectors to total health
systems, as well as from partial to integrated planning.
Originally integrated planning was dominated by the
structural-technological factors with particular emphasis
on the quantitative methods (Van der Werff, 1976).
A later development was to bring more qualitative aspects
into planningmodes, and more attention has been paid
to participation in the hope that such a planning style

could facilitate the desired organisational changes. Another
development was that the decision-making character of
planning has changed in the course of time. Complete
rationality and certainty were replaced by different forms
of flexibility and uncertainty. Those changes caused
great problems not only to the development of planning
theories, but also to the planner and to the decision-makers.
Planning developments at this moment seem to go into two
directions:

a. More attention to decision-making procedures.
   Corporate planning is a mode of planning that implicates
   a certain decision-making structure.
   It is questionable whether the health care systems in
   many countries have or will get such a structure.
b. More attention to the development of values and objectives in health care. Where do we want to go with
   health care? Among other things, scenario-methods
   seem to be possibilities to improve planning. But
   before all, this development calls for studies on
   substantial (health) planning theories.

One conclusion seems to be very clear: one specific,
every problem embracing planning method is impossible for
theoretical and practical reasons. Such a method would
also be undesirable considered from the viewpoint of
democratic and administrative aspects.
Planning replies to complexity and complexity has to do
with changing objectives, relations and structures, i.e.
complexity has to do with real social life.

7.   References

H.J. Aquina (1972): in A. Hoogerwerf (red.):
 Beleid belicht , Alphen aan de Rijn, 1972,
quotated in: A. Peper (1974).

H. Blum (1974): Planning for health; development and
application of social change theory , New York, 1974.

J. van Doorn & F. van Vught (1978): Planning; methoden
en technieken van beleidsondersteuning , Assen, 1978.

A. Etzioni (1967): Mixed-scanning; a third approach to
decision-making , in: N. Gilbert & H. Specht (1977).

A. Faludi (1976): Planning theory , Oxford, 1976.

A. Faludi, L.Th.M. Snellen, P. Thoenes & R.J. In ' Veld
(1980): Benaderingen van planning ; vier pre-adviezen
aan de WRR; 's-Gravenhage, 1980.

J.K.M. Gevels & R.J. In 't Veld (1980): reader: <u>Planning</u>
<u>als maatschappelijke vormgeving</u> , Deventer, 1980.

H.R. Van Gunsteren (1980): <u>Planning in de Verzorgings-</u>
<u>staat; van chaotisch naar systematisch falen</u> , in
J.K.M. Gevers en R.J. In 't Veld (1980).

N. Gilbert & H. Specht (1977): <u>Planning for social</u>
<u>welfare</u> , Englewood Cliffs, 1977.

E. Hall (1980): <u>Beyond culture</u> , 1977, quotated in:
Ph. Idenburg (1979).

B.M. Hudson (1979): <u>Comparison of current planning</u>
<u>theories: counterparts and contradiction</u> ; in: Journal
of the American Planning Association '45 (1979) okt. nr. 4.

Ph. Idenburg (1979): <u>Politisering van de gezondheidszorg</u> ,
in: S & D, sept. 1979, nr. 9.

W. Kickert (1979): <u>Organization and decision-making, a</u>
<u>systems-theoretical approach</u> , Amsterdam, 1979.

K. Mannheim (1940): <u>Man and society in an age of</u>
<u>reconstruction</u> , London, 1940.

M. Mastik-Sonneveldt (1981): <u>Planning over meer bestuurs-</u>
<u>lagen</u> , Nijmegen, (publication foreseen).

A. Peper (1974): <u>Bij stukjes en beetjes, over het</u>
<u>zogenaamde realisme van het incrementele beleidsmodel</u> ,
Meppel, 1974.

N.L. Rade & J. de Smit (1977): Planning, een doolhof?
In: <u>Intermediair</u>, 13, (1977), nrs. 37 and 38.

W. Salet (1980): <u>Planning theorie in perspectief</u> ,
studierapport nr. 16, Rijksplanologische Dienst,
's-Gravenhage, 1980.

J.D. Steinbrunner (1974): <u>The cybernetic theory of</u>
<u>decision</u> , quotated in: F. van Vught (1980).

P. Thoenes (1980): <u>Planning en pluriformiteit</u> , in
A. Faludi (1980).

F. van Vught (1980): <u>Planning als beleidsondersteuning</u> ,
in: Gevers & In 't Veld (1980).

R.J. In 't Veld (1980 I): Planning en democratie ,
in: A. Faludi (1980).

R.J. In 't Veld en J.K.M. Gevers (1980 II): Planning
en andere acties, een inleiding , in: Gevers en In 't
Veld (1980).

A. Wentink (1976): Sociale Planning in de verzorgings-
staat, mogelijkheden en beperkingen , 's-Gravenhage,
1976.

A. van der Werff (1976): Organizing health care systems,
a developmental approach , Eindhoven, 1976.

C. Wisemann (1979): Strategic planning in the Scottish
Health Service, a mixed-scanning approach, in: Long
Range Planning, 12,(1979) April.

MANPOWER PLANNING AND HEALTH SERVICES SYSTEMS

Gunnar Wennström

The National Board of Health and Welfare

Stockholm, Sweden

1. Background - Changing Personnel Requirements

In industrial countries the services for health and medical
care have undergone massive expansion in the decades after the
Second World War. In most countries these services now consume
7 to 10 percent of the gross national product (GNP). The number
of people they employ has also risen rapidly. Personnel costs
can be assumed to be around 50 to 70 percent of the total
expenditure on health and medical care.

In Sweden the current cost of the health and medical care
services is about 10 percent of the GNP. But these services also
employ about 10 percent of the country's total workforce.

The expansion of the services was greatest in the 1950s and
1960s. Medical progress during this period led to refinements
in diagnostic and therapeutic techniques. The problems in the
expanding health and medical care facilities were then closely
related to recruitment of trained personnel and to distribution
of personnel between various sectors of the services and geographic regions (including urban v. rural districts). The economic limitations were at that time not so serious. Measures were
taken to increase output from schools of medicine and nursing,
and from training of other staff categories, and to ensure
personnel allocation also to sparsely populated regions.
Concurrently, hospital services were increasingly concentrated
to highly specialized and differentiated units. In this process
interest became specially focused on hospital care.
Such was the background for an increased supply of doctors,

nurses and other health service personnel in many countries.
Training programmes have also tended to become highly specialized. Programmes were instituted for training of a number of
new personnel categories, with increasingly specific and
differentiated duties.

Present-day problems within the health and medical care
services in most countries are totally different. The number of
trained personnel are now relatively satisfactory, though some
imbalance remains in their geographic distribution. Indeed, the
available personnel supply continues to increase. From economic
points of view many health systems now face a situation of zero-
growth, or where overall cuts are deemed to be desirable.
Further, there is a clear striving towards reorientation and
restructuring of the systems for delivery of health care.
The main objective of this process is to provide communities
not only with medical care, but also with programmes for prevention of disease and for promotion of health. In the prevailing
opinion, the organization and content of health care must be
integrated with the community's needs as regards health and
social conditions to a much greater extent than previously.

In most industrialized countries attempts are being made to
change the direction of the health service organizations along
these lines. Increasingly there are demands for expansion of the
primary health care services and of preventive medicine and certain specialized sectors such as psychiatric out-patient facilities, care of the elderly and rehabilitation, as well as for
reduction of in-patient stay in acute somatic and psychiatric
hospital units.

In the discussions regarding manpower in the health
services, the expressions used include oversupply, overspecia-
lization, maldistribution and malutilization of personnel.
There are growing demands for a more holistic view of community
problems and for promotion of a team-work approach.
The health services have been ill-prepared for the envisaged
changes in personnel demands, which thus concern both numbers
and work content. Education and training of personnel is a long-
term project. Measures are therefore required to bring personnel
planning and training for health and medical care into line with
the longrange demands to be made on the services, i.e. to improve
health manpower planning and educational planning for personnel
within these services. It is also important, however, to consider
two other major aspects of manpower development in the health
services, viz. production and management.
Here, too, training and deployment of personnel must be improved,
as part of a programme to permit rapid recognition of altered
demands and professional adjustments to meet these demands.

The following discussion concerns some measures to secure, in the light of the above statements, a more appropriate personnel situation in the services for health and medical care.

2. Measures to improve the manpower situation in the health services

2.1. Manpower planning - an important factor in planning of health and medical care

The continuing development of the health and medical care services accentuates demands on longrange planning. In view of the financial restrictions which inevitably will characterize this development, it will be increasingly necessary to evaluate current activities, to rank priorities between various sectors of health services and to restructure activities within the framework of available resources. The necessity to reallocate resources within the services in accordance with a comprehensive policy will also considerably increase. If these needs are not observed, the risk arises that desirable changes in the services will not be feasible. For example, plans for expanding preventive medicine, primary medical care, longterm care and psychiatric services may have to be shelved.

Decisions made today are binding on the health and medical care services for a long time to come. Even absence of plans for these services implies binding to a defined course, as a rule continuation of current trends and resource allocations. Absence of planning thus, in fact, implies a form of planning, albeit an inferior form.

Longrange planning of health services must involve all activities in health and medical care - a total plan. Such a plan, however, cannot be made in the same detail and with the same precision for all components of the services. There must be a comprehensive balancing of the claims of various sectors and areas as a basis for the planning in different subsectors - for example in matters concerning manpower - on a midrange term.

A longrange view in health service planning is important not least in regard to manpower planning. At least six years are required for the basic training and qualification of a doctor. To produce a specialist, six more years are at any rate required. Change in the basic training of doctors, planning for which has been initiated, will therefore have consequences for the structure of medical practice as such only after a period of about 15 years.

Development in services for health and medical care are in
high degree dependent on availability of trained personnel.
Training of personnel thus is important for changing and improving possibilities for those in need of care and for
attaining aims set for the extent and content of such care.
Together with research and development, personnel training is
one of the primary instruments for influencing services for
health and medical care and for their continued evolution.

A redirection of the community's policy for health and
medical care services entails requirements for changes in regard
to dimensioning and content of training programmes. In work on
these schemes there must therefore be close liaison with planning for patient care, above all as regards manpower in care.
The training programmes must further be viewed as part of a more
comprehensive planning for personnel supply. At present there is
undoubtedly a considerable imbalance between the planning of
health care services as such, of manpower for these services and
of education of the involved professions.

Education for the health professions must be relevant to
the skills that will be required in various categories of personnel, and the intake for training schemes must be dimensioned
for coming needs. Educational planning thus must proceed from
the plans for evolution of health and medical care services.

The activities within a system of health care, however, are
themselves influenced by events in the sphere of education in
totally. Educational schemes designed exclusively for the health
system are neither desirable nor feasible for a progressive service. The supply of personnel at various educational levels, and
also recruitment to and acceptance for training programmes, are
dependent on the country's general educational policy.

A relative increase in the number of doctors in a country
therefore can be a reflection of a nationwide concentration on
high-level education, and not only a means of satisfying needs
in the health system. An increase in the numbers of health care
personnel with higher education is of course to be regarded as
an important investment in knowledge and skills.

The capacity for changes and readjustment to meet new
situations within the health system can reasonably be expected
to increase with the proportion of trained personnel. This
circumstance, together with the variations that may exist in
the distribution of tasks between different personnel categories,
precludes comparisons based on ratios of personnel - e.g. of
doctors to population by year - as an adequate instrument for
comparisons of health and medical care schemes within a country
at different times and between different countries. For the same

reasons, it is difficult to use such ratios for describing
other aspects of the services, including deficit or surplus
of doctors.

The doctor : population ratio is thus an expression of a
country's policy for health and medical care, which includes
allocation of duties between different personnel categories,
but also a consequence of the country's general educational
policy. The sex composition of health systems staffs is likewise
influenced by these circumstances, as is the willingness of personnel to be employed in work for which they have been trained
(proportion of trained persons who are actively engaged in their
professions and their degree of activity, calculated as hours of
work per week, month or year).

Questions of "oversupply" or "surplus production" of any
specific category of health service personnel therefore must not
be considered in isolation from other categories. These
questions must be placed in a wider context and viewed as
problems of balance in relation to the tasks required in the
system and the possibilities for deployment and financing of
personnel. The balance concerns also distribution of tasks
between personnel categories and professional competence of
existing personnel to meet demands made on health service sectors. Other aspects are distribution of the system's personnel
resources geographically and between different sectors and
specialties of health and medical care.

2.2. A comprehensive policy for health service manpower

A prerequisite for the varied activities concerning developments in manpower for the health services is a solid
basis to which all matters connected with the complex problems
of supply and demand are to be related. There is a need for
formulation of a comprehensive policy for health service personnel as an integrated part of a national policy for health and
medical care. This policy must be soundly constructed and must
achieve a solid political status aimed to promote fulfilment of
the nation's overall policy for health care.

The concept comprehensive is used here in accordance with
recommendations from WHO on health manpower planning, which
imply influence on as many variables as possible, both qualita-
tively and quantitatively, based on existing policy for health
service manpower. Systems for health care and for education of
the involved professional groups are highly complex. Numerous
parallel activities on different levels must be planned, implemented, monitored and evaluated in a continuing process.

Objectives for the professional education and training
of the health service personnel groups must also be formulated
in such a comprehensive manpower policy. Some of the requirements for qualitative changes in training programmes that may
then be suitable for inclusion in an educational policy are
listed below.

- increased direction of the health services towards preventive
  measures, primary health care and care of the elderly;

- heightened interest in a holistic approach to human problems;

- as a consequence, attempts to restrict specialization, with
  base level specialization related chiefly to the problems of
  primary health and general hospital care;

- co-ordination of training programmes for different personnel
  categories with, as far as possible, common basic training
  and thereafter graduated differentiation - present-day
  personnel and educational structure within the system has
  been criticized as excessively fragmented;

- closer collaboration between different personnel categories
  (team care) in health and medical care, and between this
  system and related sectors - primarily the social services,
  and

- redistribution of duties between personnel categories, the
  considerations including the changes in personnel supply
  that are to be anticipated as a result of the developmental
  work being done in various training programmes.

The listed objectives necessitate well developed collaboration between the health planning sector, including manpower
planning projects, and the planning of education for the health
professions. The forms for such collaboration will of course be
governed by the circumstances prevailing in the respective
countries.

2.3. Optimum utilization of existing resources

Because of restrictions on expenditure for development of
health and medical care services, effective utilization of
currently available resources is essential, and also adaptation
of these resources to meet changed demands.
Experience has shown that it is not possible to simplify and
increase efficiency in health service activities by replacing
people with machines in the same way as in industrial processes.
Medical and technical advances in diagnosis and treatment gene-

rally require additionally specialized human contributions.
Present-day health services are sometimes criticized for too
ambitious use of technique to improve efficiency. For example,
elderly persons and others often complain of difficulty in maintaining contact with the same doctors and nurses as on previous
occasions. A team-work approach in health and medical care has
as one of its objectives to counteract these disadvantages.

I have previously mentioned the long time that often must
elapse before the effects of changes in basic training are
reflected in the activities in health and medical care. Thus
as regards doctors, around 15 years may be required before a
specialist, fully trained, can put into practice changes that
were introduced in basic training. For changes in postgraduate
medical education (specialist training), the corresponding
period is five to seven years.

The quickest effects are obtained by continuing education
of health personnel. This educational form must therefore be
promoted as an instrument by which improvements in knowledge,
skill and attitudes can be achieved more rapidly than by changes
in basic or specialist education. Continuing education, moreover,
can encompass personnel in the health services at any time in
their professional life.
Schemes can be directed towards key categories of personnel, for
instance physicians, or to composite key groups such as staff
in primary health care. Continuing education thus is one of the
optimum instruments for adapting health and medical care to
altered demands, including reorganization and restructuring
activities within the services.

3.   Concluding remarks

Health and medical services in most industrialized countries are now encountering an entirely new development phase.
Demands for restraint in the pace of expansion - down to zero-
growth or even allround diminuition - and altered policies, with
more emphasis on preventive measures, primary health care, etc.,
will necessarily involve reorganization and restructuring of the
various health service activities to meet the 1980s and 1990s.
Manpower planning and planning of education for the professions
in health and medical care are the most important instruments
in this respect.

Longrange planning faces heightened demands in the current
situation for health service systems. Educational planning for
the involved professions must take place in close cooperation
with health service planning - above all that concerning its
manpower.

The content of educational programmes for the health professions must fit the demands on knowledge and skill from the
respective categories of personnel. Intakes for such programmes
must be related to prospective personnel requirements. In these
matters, however, consideration must be given to the influence
of a more general educational policy on the future supply of
personnel with different levels of education.

There is a need for formulation of a comprehensive policy
for manpower in the health services, as an integral component
in a national policy for health and medical care. In accordance
with such a policy, objectives for education of professional
categories in the services must also be formulated.

One of the most important instruments for optimum utilization of existing resources is continuing education, which can
serve also to reorient activities in health and medical care
in response to altered demands. Included among these demands
will be reorganization and restructuring of the work of the
services.

FINANCING AND HEALTH SERVICES SYSTEMS

L. Delesie

Centrum voor Ziekenhuiswetenschap

Leuven University, Belgium

Introduction

The health services are a system with many elements and a
myriad of relationships. This point has sufficiently been stressed
in the literature. Consumers, patients, physicians, institutions
and regulators interact with each other and also with the society
at large. Indeed one should avoid to look at the health services
as a closed system. The recent public outcry about the health
care cost explosion for instance is equally influenced by the
abruptly limited growth figures in the other sectors of economic
life as by the health care sector figures as such. The meager
results of hospital planning efforts in many western countries
are not the least due to the fact that the agencies involved did
not sufficiently take into account the high status and attractive
employment opportunities of a hospital within its local community.

Finances are just one element in the framework of the health
system. It is though an element most readily understood by the
public at large and its political bodies. It is also the greatest
common denominator available yet to look into the many wildly
divergent aspects of the health system such as the impressive
look of its centers of activity, the sophistication of its equipment, the status and income of its care providers, the level of
services rendered and the cost to society of this health care
industry.

This paper will necessarily have to be incomplete. The temp-
tation to touch on the myriad of driving forces which steer the
health services from within and from the outside in each country
is great. A systems analyst may validly agree that one cannot do

without. However the state of the art from a theory point of view is still rather limited at both sides of the Atlantic. Scientific results are even more sketchy notwithstanding the impressive findings already at hand. Hence, rather than skimming over many issues, we will deliberately expand more thoroughly on a very limited subset of the problem area at hand. This paper concentrates for illustrative purposes on hospital finances.

During the periode 1945-1965, the health care system in most western countries was characterized by a predominance of hospital affairs and hospital development. Post war reconstruction, advances in hospital based medicine, the powerful position of the hospital-based medical profession and of hospital-related academics, the link-pin position of hospitals in an otherwise dispersed field of care providers reinforced a hospital-centered vision of the health care system (1).

Today, the predominant position of hospitals is fortunately under investigation. The economic constraints of recent years though have still focussed the public's- and the politicians' attention overwhelmingly towards hospitals as the prime target for cost containment. Proposals, regulations, laws, experiments are in the works or already operating in many western countries. Hence hospitals financing offers a more valuable starting point for systems analysis.

We will highlight the systems effects of different alternative financing mechanisms and aim at some guidelines of interest in the context of finances in the health services.

A lengthy description of the different hospital systems and the different alternative finances experiments is beyond the scope of this document. This paper presupposes a reasonable familiarity with the health care systems and the hospital finances experiments in the western countries. By now, the international literature has published adequate background material.

In a first paragraph we will structure our approach and define the concepts which are introduced later on. In a second paragraph some common characteristics are presented. In a third paragraph we will make a tentative appraisal: tentative as few experiments have yet had sufficient test time. The major part of this document is based upon extensive research work and site visits of the Center for Health Services Research of the Catholic University of Leuven and is published elsewhere (2).

Definitions of Concepts - Structure of Approach - General Survey

Finances is still too broad a concept to be of use in this paper. We differentiate between costs and reimbursements. Costs can be either real (historic) or budgetted. One should keep in mind that even generally accepted accounting procedures still only approximate real costs. Discrepancies are particularly enhanced when many different payers interfere such as different national agencies: subsidizing agencies versus rate agencies.

Costs are fully or partially reimbursed by the different particular financing mechanisms. These are the main focus of this paper. Financing mechanisms are either open market (e.g. supplementary charges, private room physician fees, non-reimbursable drugs) or regulated. When they are regulated other objectives start intervening besides paying for costs and may become dominant.

The research project identified and investigated seven objectives (3):
1. Improving the supply of health care resources,
2. Improving the accessibility to health care,
3. Improving the organization of the health care system,
4. Improving the efficiency of health care production processes,
5. Improving health care management information,
6. Simplifying administrative procedures in the health care sector,
7. Improving the quality of care delivered.

In the context of these objectives and with a view to a systems analysis, hospital care financing mechanisms can be measured in function of their reliance on the production of services or output. Traditional fee-for-service systems, or per-diem-rate systems do stress output of services or in-patient days. On the other side of the spectrum, a lumpsum financing mechanism may completely disregard output. Real-life hardly provides examples of clear-cut situations. A plain fee-for-service system will eventually discount shifting cost patterns due to e.g. automation. A lumpsum payment provides for special treatment of some cost elements. E.g. the Swedish hospital budget procedures allow that budget allocations for pharmaceutical drugs can be overrun without any prior notice. Most real life financing mechanisms offer a mixture of output and non-output oriented elements.

Table 1 gives a survey of some financing mechanisms with regard to their emphasis output.

A lengthy analysis of each financing mechanism is beyond the scope of this paper. A too short outline necessarily incomplete from the systems point of view though, will help the reader along and hopefully arouse his interest in further reading.

Prix de journée eclaté - France

This experiment in (originally) some 3 french hospitals was rushed through in the first half of 1977. Basically an "à la carte" billing substitutes a traditional "menu" approach (one single

TABLE I

Some financing mechanisms classified with
respect to their emphasis on output

Fully output oriented

1. Prix de journée éclaté
(itemized per diem rate)
France

2. Artikel 9 - Previsionele
Prijs 1981
(prospective rate - 1981)
Belgium

Partially output
oriented

3. Previsionele Prijs 1982
(prospective rate - 1982)
Belgium

4. Réforme de la tarification
(new tariff proposals)
Hospital Federation of France

5. Diagnostic Related Groups
Costing - New Jersey

Limited output
oriented

6. Maxicaps - New York

7. Budget Global - Quebec

8. Hospital Budgets - Sweden

9. Budget Global - France

10. Experiments on budgetting -
The Netherlands
----

11. Resource Allocation
Working Party - England *

---

* The RAWP proposals are treated separately as they favor the
use of non-hospital related characteristics to determine
hospital budgets in a relative way.
Standardized mortality rates, age-sex-ratios in the region,
etc.,will determine the regional hospital budget.

per diem rate for all inpatient days in the hospital). It was
indeed hypothesized that a much more detailed itemized bill
closely following the actual production of small individual
services for any given patient would induce the hospital as
well as the patients to make more efficient use of resources
and hence result in cost savings. It is of interest to note
that this financing mechanism did raise the issue of equity as
some patients have to pay 30 times the amount which they would
have been billed under the old mechanism.

Article 9 - 1981 and 1982 - Belgium

The traditional hospital financing mechanism in health
insurance countries including Belgium (Article 9 - 1981) is the
average per diem rate for a given production "batch" of in-
patient days (usually a full year). Input elements (e.g. personnel) or cost elements (e.g. food bills) are negotiated or
regulated (growth rates, ceilings, etc...) for the batch as a
whole. Higher service levels are remunerated only in as far as
they result in more patient days and conversely. This mechanism
holds a strong incentive to steadily increase the number of
patient days offered in the market place. Little concern is given
to other considerations such as type of services, level of services, efficiency, etc... The incentive may be weakened if additional patient days beyond the given batch are remunerated at
a lower rate. This is what the article 9 - 1982 financing mechanism in Belgium does. If the marginal cost to the hospital of
additional patient days is less than the marginal per diem rate
(in casu 60% of the normal per diem rate), the hospital has less
advantage at producing those extra patient days and may redirect
its priorities (in casu a more rigorous admission policy). A
drawback of the "batch" notion is that it somehow institutionalizes
historic inefficiencies.

Réforme de la tarification - France

In 1980 the French Hospital Federation developed its counter-
proposals to the experiments run by the government. In a nutshell
the Federation favors a non production related base remuneration
(budget primitif). However this base remuneration is reviewed
and finalized from year to year on the basis of production figures
(inpatient days and number of admissions) which are applied to
the variable costs only. This "compromise" proposal has not been
put to test yet.

Diagnostic Related Groups Costing

In 1980 New Jersey started to pay 26 hospitals in a different
way. Essentially, fixed costs with respect to patient care (e.g.
plant depreciation) are budgetted in some, sometimes arbitrary way

(e.g. 20 $ per resident per discharged patient). Patient care
costs are remunerated on the basis of a fee per patient of a
particular category. 383 patient categories are identified on
the basis of diagnosis, surgery, age, sex, service and length
of stay. The fee per category is arrived at through analysis
of case mix and cost figures among the group of 26 hospitals.
This mechanism stresses the production mix and the production
function rather than the production level. The hospital care
process becomes dominant.

Maxicaps - USA

In 1980 an agreement was arrived at between 24 hospitals,
insurance agencies, the planning agencies, the consumers and the
regulating agencies in the Rochester Area, New York, to implement
a new financing mechanism. Basically this mechanism is a huge
(several hundreds of pages) contract among all parties involved
where the revenue implications of all production eventualities
are a priori negotiated. Capital expenditure is dealt with separately from running costs. Inflationnary trends, wage cost trends,
emergency situations such as plane crash accidents, contingency
funds are all covered in the huge contract for the year to come.
Maxicap is the ultimate in "no surprises" hospital financing.

Budget Global - Quebec

The Quebec budget global was introduced 10 years back. Essentially, government imposed stringent levels of overall hospital
expenditure with a view to reallocation of resources towards
primary care and chronic care. The yearly budget adjustments
posed the main challenge. These were worked out on the basis of
stringent cost control (e.g. less than inflationnary trends) and
productivity ratios rather than production figures. The productivity ratios are arrived at by comparison among similar hospitals.
Similar hospitals are defined in a cluster analysis which takes
into account as well the structural elements of the hospitals
(teaching staff, special services, etc...) as their diagnostic
mix.

Hospital Budgets - Sweden

The hospital budgets in Sweden are the regular budget financing
mechanism. However, the constraints and the organizational set-up
within which the budgets are arrived at are of interest. The constraints are basically the three levels of plans: long term plans:
15 years, short term plans of a rolling off, rolling on type -
called RUPRO - and yearly plans finalized in the budgets. The
organizational set-up to implement and elaborate the plans including
the yearly budgets, has been experimented on. In 1977, a block

structure was introduced which grouped similar activity centers within a district but across hospitals. E.g. all internal medicine departments together had to produce a common consensus proposal for the district-authorities. As of today however the block structure is planned to have an advising role only.

Budget Global - France

Parallel to the "Prix de journée eclaté", France also introduced a "budget global" experiment to find new ways to improve the chronic cash glow disease prevalent in french hospitals and to increase physician involvement in hospital affairs. Though production figures do not determine hospital revenue in a direct way, they are still the linkpin of the budget. However production is measured in an entirely new enlarged way. Inpatient-days data are argumented by not only case mix data - cfr. DRG Costing in New Jersey - but also by therapeutic activity data including the economic impact of different therapeutic alternatives upon the hospital. An increased outpatient activity to the detriment of inpatient therapeutic care has already been shown.

Budgetting Experiments - The Netherlands

By the beginning of 1982 a three-years experiment was started in some 4 dutch hospitals to substitute the traditional fee-for-service financing mechanism by a budget financing mechanism. The development of new managerial ways to control the hospital care production process is one of the prime objectives. Better control, improved quality of care, higher efficiency, better information systems are anticipated and will be closely monitored during the experiments.

Resource Allocation Working Party - England

This financing mechanism, partially implemented since 1976, reallocated DHSS financial resources among different health care regions and their hospitals based upon less hospital-related criteria such as age and sex adjusted incidence rates, by medical specialty, each specialty having its cost weight factor. It is evident that the equity issue plays an important role.

The financing mechanism is the linkpin for the reimbursement of costs. Once determined, several other system characteristics still have to be decided upon, which are however - it should be stressed - essentially independent of the financing mechanism selected.

For instance all mechanisms can be prospective as well as retrospective. Again a mixture is usually prevalent. Some retro-

spective elements being grafted on an essentially prospective
mechanism, or the other way around. A second characteristic is
the strategy used for cost control. Two main strategies and their
concurrent sets of criteria can be identified (4). The first can
be called the formula method. Individual items such as wage rates,
occupancy rates, staff ratios, depreciation rates are subjected
to specific formulas which will limit cost developments for some
cost centers in some hospitals but by the sum taken will set goals
for other cost centers in other hospitals. The review method is
more all-encompassing hence allowing for individual characteristics
to be taken into account. Again a combination of both is usually
the case.

A third system variable to be set is the payment mechanism
for the reimbursements. In a national health service these payments will most often be direct on the basis of monthly accounting
statements. In a national health insurance system a network of
middlemen - third parties or insurance agencies - services the
needs of its clientele. Though their activities are overwhelmingly
plain administrative, their role yields usually much more power
from a system point of view: they operate on the micro-level of
the health care system and, through their financial leverage they
are strategically located to exert some control on the decision-
makers, the physicians.

An interesting subcharacteristic of the financing and/of payment mechanism is the degree of co-payment, co-insurance, supple-
mentation, prior approval and referral conditions, etc. for some
or all types of services and their effect either on production for
specific kinds of providers or demand for specific population
target groups. Many hypotheses have been formulated, many emotional
statements have been made but evidence is still scanty (5).

Though the basic concepts with respect to finances in the
health system are rather straightforward real progress is extremely
slow:

- finances are instrumental in the pursuit of other objectives
  than the reimbursement of costs (see above);

- societal viewpoints, historic and policy developments hardly ever
  push a health system unidirectionally towards a clear-cut situation. Our research always identified large or small offshoots
  of alien financing mechanisms in the indigenous financing mechanism (6);

- little comprehensive systems understanding is available - and
  indeed hardly ever published - about the micro-level of the health
  care system. The lack of data and the high cost of gathering these
  has resulted in an emphasis on macro systems research in the
  health field, in any case with respect to finances.

Salient Features Common to the Different Financing Mechanisms

Systems research has kind of a predisposition to look for differences among systems. We do esteem that the common features will help more to understand finances in the health care system than the myriad of interesting and/or trivial differences.

1. Parallel Situations

Our research revealed more common traits than points of differences. Moreover differences are primarily due to different traditions and administrative developments and timelags in the evolutionary trends of the countries under investigation. It should be noted that no financing system claims to be the golden bullet. Most mechanisms are experimental, temporary, transitional or explicitly state the need for further refinements.

2. Finances and Health Care Policy

Beyond doubt, the main policy issue at this moment in time in the health care sector in the western countries is control. More often than not, finances are called in to countervail the control instruments of a previous age such as certificates of needs, planning and programming or registration procedures, PSRO, profiling efforts, and subsidy policies which seem to be gone broke vis-à-vis the main control issue as of now: the control of costs. Contradictory situations are still abundant. E.g. norm setting efforts on the one hand are still dealing with developing minimum standards while finances are hurriedly called in to fix maximum ceilings for expenditure. This type of conflict is particularly apparent when investment or structural programs (input programs) are compared with running or operational programs (process, output and outcome programs).

Systems researchers may find here an attractive area of application.

Finances are also called in to deemphasize the hospital sector within the context of the total health care system. Few efforts so far have been able to reach beyond the stage of limiting hospital growth rates.

3. Finances and Decreasing Hospital Costs

All financing mechanisms have clearly stated that they want to decrease the total level of resources available to the hospital sector.

Moreover finances as a system variable seem to be the only tool available to obtain this goal, other efforts such as bed programming or manpower planning being insufficiently successful.

A curious trend is developing: on the one hand government initiative is on the increase, on the other hand the same government initiative wants to increase the level of financial responsibility and accountability in the individual operating units.

Two driving forces enhance this apparently contradictory trend. First, the development towards further regionalization of the health care sector is reaching the stage where finances - the ultimate bastion of central government control - are also beginning to be regionalized. Second, the development towards further democratization is expanding the power basis in the individual hospitals to include also employees, physicians, local politicians and occasionally, consumers.

4. Finances and the Transparency of the Hospital Sector

A thorough analysis of the different hospital financing mechanisms also reveals that finances have been instrumental in rendering the hospital system more transparent, hence susceptible to more control.

Four evolutionary stages can be identified.

The first stage, already gone by in most western countries, we call the reimbursement stage. Finances were only instrumental in providing the hospitals for the means of income necessary to run their services. Few questions were asked. Tariffs, per diem rates were periodically revised according to some national guidelines such as wage indexes, inflationary trends.

The second stage we call the cost stage. As questions were beginning to be asked by the controlling agencies, the financing mechanism was used to identify and measure detailed cost data by cost center and by type of cost. Uniform accounting schemes were introduced; pricing procedures, pricing standards, formulas, cost limits were worked out. Rates and budgets were reviewed including, excluding or limiting specific expenditures.

This cost stage however did not provide for sufficient information to control the hospital process. Hence the cost stage is followed by, what we call, production stage. Costs are beginning to be looked at in a relative rather than an absolute sense. Costs in relation to the average length of stay or occupancy rate. Costs in relation to the number of nurses per patient, the number of surgical procedures, the number of new admitted patients per day, costs per square feet, etc... The production stage indeed concentrates on the hospital efficiency. "Expensive" hospitals are penalized, "cheap" hospitals get a bonus. Economies of scale are reintroduced through the concept of multi-units.

Unfortunately, the production stage overlooks what happens eventually to the patients or the hospital outcomes.

A fourth stage, which we call the patient profile stage is emerging. DRG's and other case mix or hospital classification schemes are being introduced to evaluate production in view of hospital outcomes.

Financing mechanisms become instrumental in cost/benefit analysis or even zero base budgetting.

Finances are being integrated in a more complex information system, the purpose of which is to render the hospital sector more transparent in many dimensions besides costs in order to improve effective control.

5. Finances: Supply or Demand?

A last issue with respect to finances in the health care sector is the demand-supply issue. As of now all finances experiments are overwhelmingly geared towards the supply side. (Belgium is one of the few countries where the demand side is also dealt with). A special topic within this context concerns the so-called profit issue or carrot motive. Are profits due to the efficiency of the daily activities or do they result out of inferior quality of service? Should profits stay within the institution or should they be spread among different institutions? To what goals can profits be applied?

As of now the different financing experiments do offer few answers, maybe due to the fact that the situation itself is rather exceptional at the moment.

An Appraisal of the Differences Among the Financing Mechanisms

A thorough discussion of the many hypotheses which underly the different financing mechanisms requires volumes. The main issues though from a systems point of view will be introduced.

Financing mechanisms which are primarily production based- e.g. fee-for-service - do highlight hospital costs whether they are formula oriented or favor a lumpsum strategy. Non-production based financing mechanisms raise the issues of quality of care and of hospital effectiveness from an organizational as well as patient care point of view. These issues are tougher to decide upon. Hence a tendency can be observed to stick to a middle-of-the-road position. Governmental agencies as well as hospital administrators are looking for ways to deemphasize production while

avoiding to have to take a stand on outcome of diagnostic and
therapeutic activities.

A production based financing mechanism more easily allows
hospital planners to plan alongside or independently of norm setting
agencies and financing agencies. A non-production based financing
mechanism demands for a more global approach which may result in
cumbersome conflicts and stalemate further progress. These remarks
hold for the hospital itself (administrators versus physicians) as
for the relationships between the hospital and the regulating
government.

Indeed a more global approach requires a different, new kind
of information on top of the available data. This type of patient-
oriented and quality-of-care oriented information is not yet
readily available for day-to-day managerial purposes.

A production based financing mechanism brings about that a
higher level of production instantaneously results in more income.
This is not so in a non-production based financing mechanism. The
first approach is favored by the institution because of its fast
feedback and by the controlling agencies because of its straight-
forwardness in the annual budget process.

The non-production based financing mechanism complicates
matters considerably but allows on the other hand for long-term,
pinpointed managerial decision-making. The reverse of the medal
however is that in times of financial hardship the institution,
strait jacketed by the lumpsum, maybe tempted to start favoring
primary care hospitalizations over secundary or tertiary care
hospitalizations or will decrease its level of services. Waiting
lists and poor service may be the result.

A positive side-effect of the non-production based financing
mechanism is that the institution itself is not interested in more
hospitalisations, lengths of stay as such and may even take initiative to reorient its resources itself towards other types of care
such as ambulatory care.

A production-based financing mechanism is more often associated
with inflated lenghts of stay, unwarrantable hospitalizations, and
overproduction in general (7). This built-in reaction is particularly
likely when an oversupply of hospital care facilities is emerging,
as many western countries are discovering (8).

A drawback of a non-production-based financing mechanism
regularly cited is its inherent tendency towards stagnation. As
the second preceding paragraph identifies a reverse motive, some
explanation is due. Indeed as the hospital sector is rather labour-
intensive, it is most likely that the lumpsum allocation will have

TABLE II

Distinguishing features of a production based
versus a non-production based financing mechanism.

REIMBURSEMENT CRITERIA

cost data--------------cost and----------------cost
                    productivity data   productivity
                                             and case mix data

REIMBURSEMENT STRATEGY

annual	------	long term
individualized per hospital	------	for a group of similar hospitals
averaged over the different services	------	individualized per type of service

REIMBURSEMENT CONTROL

complete annual revision	------	marginally updated allocation
annual budget control	------	long term budget control
cost ceilings	------	variations allowed within limits
itemized mandatory growth rates	------	internal adjustments allowed

REIMBURSEMENT MECHANISM

per patient day per itemized service	------	yearly basis
proportional to production	------	allotments
detailed and itemized bills	------	summary lists

PRODUCTION BASED	------	NON-PRODUCTION BASED

(cont'd)

TABLE II (cont'd)

IMPACT OF MANAGEMENT

administrators and physicians work alongside each other	------	consultation and cooperation between administrators and physicians
planning and financing independent of each other	------	cost benefit evaluation of projects
emphasis on budget control	------	control of the total hospital

PRODUCTION AND PRODUCTIVITY

lower treshold for hospitalization	------	selection of cases
good service	------	waiting lists, poor service
incentive towards greater efficiency	------	stagnation
inflated lengths of stay	------	cheap hospitalization
overemphasis on hospitalization	------	room for new initiatives

---

PRODUCTION BASED	------	NON-PRODUCTION BASED

to rely in part on manpower criteria and their wage-agreements. These are well known to resist any change in hospital policy. This issue of stagnation versus incentive towards initiative is a good example to illustrate the seemingly conflicting nature of the hypotheses which we are yet dealing with, with respect to systems analysis in the hospital sector. Future research still has a lot of opportunities.

Table II which summarizes some of the drawbacks and distinguishing features of the two divergent financing mechanisms, may help in directing future efforts.

Conclusion

This document demonstrates the importance of finances in the health care system, more specifically the hospital sector. Finances being one instrument of health policy, influence resource development, resource allocation and resource utilization. The few examples given though, show that finances are not a panacea. Some salient trends in the health care sector follow their course independent of the financing mechanism. On the other hand, the financing mechanism is able to steer the hospital sector into different directions. Several finances "variables and relationships" have been identified. Many more have not been reported on. As of now knowledge about their impact on the hospital care sector is still very much in the hypothesis stage. A lot of research and developmental work on the micro-level still has to be done. It is indeed expedient as of now to call any new financing mechanism experimental.

References

1. J. Blanpain, L. Delesie, H. Nys, "National Health Insurance and Health Resources", Harvard University Press, Cambridge, 1978.
2. L. Delesie, R. Debackere, J. Deridder, J. Leenders en H. Spinnewyn, "Alternatieve financieringswijzen voor de werkingskosten van Ziekenhuizen", Ruhamco, Leuven, 1980, 308 p.
3. Other objectives have been come across lately e.g. the Dutch budgetting experiments.
4. W. Schwartz, "The regulation strategy for controlling hospital costs", N. Engl. J. Med. ,1981, Vol. 305, p. 1249.
5. J. Newhouse, "Insurance Benefits out of a pocket payment and the demand for medical care : a review of the literature", Rand Corporation, Santa Monica, 1978 and subsequent publications.

6. See also A. Culyer, A. Maynard and A. Williams, "Alternative systems of health care provision: An essay on motes and beams, in M. Olson (ed.) New Approaches to the Economy of Medical Care", American Enterprise Institute Washington, 1982, p. 131.

7. D. Banta, "The diffusion of the Computed Tomography Scanner in the V.S.", Int. J. Health Services, 1980, Vol. 10, p. 251.

8. J. Bunker, B. Barnes, F. Mosteller, "Costs, Risks and Benefits of Surgery", New York University Press, 1977.

COST-EFFECTIVENESS AND HEALTH SERVICES SYSTEMS —

A SYSTEMS APPROACH TO THE ASSESSMENT OF HEALTH CARE

Charles O. Pannenborg

Ministry of Health, The Hague

The Netherlands

1. Introduction

The terms cost-effectiveness and systems-analysis are relatively new to the Western world of health services. Both concepts originated in military operations research, made their way successfully through big industry and its corporate planning, and from there on now slowly penetrate the health sector's management and planning.

While the development of the two notions within the health sector runs almost parallel in time, neither so far has a bearing on the other; in other words, despite their common heritage, cost-effectiveness and systems-analysis of health services remain rather oblivious of each other. Indeed, whereas their scientific basis and explanatory capacity is increasingly convincing and powerful in terms of overall health policy, their practical relevance for day-to-day medical activity, hospital management and patients' needs seems, however, to develop in ever more isolation.

In order to stimulate the prevention of the progression of ad-hoc experimentation and random selection of problem-solving, we propose to clarify the value of linking the two concepts of cost-effectiveness on the one hand and systems-analysis on the other together; doing so may render the development of both of them in the health sector more successful. Considering the specific nature and properties of each of the two topics, it is clear, of course, that the development of cost-effectiveness with respect to health services will benefit more from the application of systems-analysis than the other way around.

The issue of assessment is growing increasingly important
in the health sector of most countries where recession has
deflated the spurious ideas of an open-ended welfare system.
Cost-effectiveness being one of the major potential instruments
of the assessment of health services, it seems valid to highlight health services evaluation as one of the major components in any future comprehensive system of health services.

2. Definitions and Terminology

The definitions of "effectiveness" and "cost-effectiveness"
in the field of health care, unfortunately, have become a problem-
atic matter almost all over the world. Most managers and national
health officials now are familiar with the term. Chances are,
however, that in by far the majority of cases different concep-
tualizations and deviating understandings occur between and among
those who use the terms in their discussions on, for example,
health policy, hospital economics, or screening in preventive
medicine.

The most important distinction is to be made between cost-
effectiveness analysis (CEA) and cost-benefit-analysis (CBA).
Although this is not the proper place to expound the differences
between CEA and CBA extensively, we need to clarify at least
some major systematic variations.

CEA is by definition an ex-post instrument of evaluation.
CBA is by definition an ex-ante policy instrument. CBA is an instrument to choose between different options and objectives before
starting the activity, project or programme, whereas CEA refers
to the choice between the various effects of one of the policies
concerned. Thus both are techniques to assess, beforehand or
afterwards, the negative and positive consequences of alternative
procedures. (1)

The major objective of CBA is to value the desired results
or benefits of a health policy or programme; they are valued
in numerical units, e.g. in terms of money, in terms of costs.
This allows to compare the benefits and the costs and to determine
whether the benefits exceed the costs. It equally implies the
more fundamental assessment whether, in view of the nature of
the to be expected benefits, it is worth making the costs at all.
This holds especially for CBA, as the numerical valuation of both
the benefits and the costs refer to the quantification of such
diverse issues as "the social benefits" or "the social costs" of
elderly human health or chronic disease, the "time-cost" of training and educating medical manpower, the capital- and (un)employ-
ment-costs of hospital or medical school construction, the
"opportunity-costs" of launching certain vaccination programmes
at a given moment in time, the "compliance-costs" of certain

additional diagnostic services, the "income-political" and "multiplier" costs of different health practitioner remuneration systems, the "out-of-pocket" costs in budgetary hospital management, etcetera.* Thus, CBA asks for the determination of the largest benefits versus the costs of various competing programme alternatives. The programmes can be of widely divergent types, for example, whether certain investments should be made for the construction of new traffic facilities or for the expansion of an intensive-care unit.

Cost-effectiveness analysis, on the other hand, attempts to measure the output of an activity not in monetary terms but in some other unit, such as, for instance, potential years of life saved, days of morbidity or disability avoided or number of diagnostic procedures diminished. CEA starts with the defined, supposedly desired, objective (the to be attained effect) of a certain activity, programme or project. It does not question the validity or the inherent worth of the objective, nor is it able to compare programmes of a different nature.

CEA does question whether and to what extent the desired effect can or has been attained and at what costs. Secondly it questions whether the same or a better effect (e.g. a higher no-recurrence rate or an improved 5-year survival rate) could have been reached with other means within the same programme. It is at this point that confusion often occurs, since in health care and in medicine the capacity to define the exact status of the objective to be reached, i.e. of the desired effect on the medical procedures envisaged, generally is ill-developed. In other words, the norm or standard to be attained, often remains wooly, with the consequence of a disproportionate emphasis of the costs involved. Conceptually cost-effectiveness should be divided in the issue of effectiveness on the one hand and cost-effectiveness on the other. Returning to the issue of norms and standards infra, at this stage it is sufficient to note that the first major impact of the introduction of CEA into health (services) systems is the growing awareness and increasing attempts to define specifically for each health and medical activity what effect or effects they are meant to

---

* Opportunity-costs, for example, are the costs of a hypothetical activity that would have been possible instead of the activity which is actually chosen to reach the objective. They are the costs which accrue because the one activity is done instead of the possible other, thus giving up the opportunity to do the possible other: Robinson Crusoë pays nobody, but realizes that the costs of his picking strawberries are, for instance , those of the sacrified raspberries which he had the opportunity to pick at the same time and with the same effort. (2)

produce. This clarification of objectives, beyond such goals
as "curing the patient", "relief of suffering", etc., would
enhance the ability to manage health and medical programmes
enormuously.

The second major pay-off of CEA then should be the introduction of a mode of thinking which weighs and analyzes all
the possible alternatives. Apart from differential diagnoses,
the health and medical professions experience large difficulties
in conceptualising a number of different, all possibly effective,
procedures to be executed, to weigh them and then to select
the most appropriate one for attaining the highest ratio of
success of the predefined objective. Defining 100%-effectiveness
under the most ideal circumstances with the term "efficacy",
effectiveness always constitutes a ratio. The degree of success
in reaching the, preferably quantified, predefined objective of,
for example, an acute frontalis, could then in very sophisticated
surroundings and with very high skills, such as to be found in
some health science centres, approach 90 or 95%, whereas the
majority of peripheral surgery-outcomes would have to do with
65 or 70% effectiveness.

The issue of effectiveness vs. cost-effectiveness in terms
of differential procedures relating to outcome can well be
illustrated by the example of cervicodynia. At present radiodiagnostics generally does not have a quantitative insight
in the results of its procedures with respect to patients with
prolonged cervical pain. The assumption is that something like
80, 85, 90 or perhaps 95% of all cases regards degenerative
afflictions, for which there are no effective therapeutics.
20, 15, 10 or perhaps 5% of cases, however, may show a tumor.
The effectiveness of the radio-diagnosis, then, depends on the
question whether and to what extent the effect of the tumor-
therapy changes when radiodiagnosis is deferred to and only
performed at a later moment when the indication of a tumor has
become tumor-specific. Or, to phrase it differently, whether a
reduction of radiodiagnostics regarding cervical pain (being
ineffective for degenerative cases) will make any difference
for the (choice and effect of) treatment of the tumor-cases that
will be diagnosed in a later stage anyway.

Cost-effectiveness comes to play a role from the moment
when the costs of radiodiagnostics and their decrease when
abolished for degenerative cases, is put versus the negative
or positive costs of the tumor-diagnostics and treatment. All
this, of course, within the margin of the value-judgements
whether a possible negative change in outcome of tumor-treat-
ments is acceptable or not; i.e. whether a 5% decrease in treatment effectiveness justifies the dramatic increase of the (negative) effectiveness of the diagnosis of degenerative cases.

The 5% loss, then, should equally be compared with the long-term effectiveness of the treatment.

At this stage CEA moves into the field of risk-effectiveness analysis, where the desirable outcomes are put against the undesirable consequences, such as nosocomial and iatrogenic cases. Thus, in risk-effectiveness analysis, the potential of orthopaedic surgery to prolong life or to avoid disability is compared with its operative mortality and postoperative morbidity.

3. Evaluation of Health Services

Over the last twenty years, we have witnessed an ever-growing body of evaluative ways and means in the health sector. The largest thrust came from the industrial and behavioural sciences that invented and refined the evaluative techniques and demonstrated clearly their usefulness and necessity.

The second thrust came from the conglomerate of interests that were concerned with the quality of health and medical care, and which, in view of the fast growing scope and potentials of the health sector and of the science of medicine, were actively in pursuit of mechanisms that somehow could help in defining and measuring quality of care. Among these interests we find the medico-legal societies, the health insurance companies and the regional and national health administrations including their Inspectorates and Quality Control Boards.

The third thrust came from the various professional societies and associations in clinical medicine. A growing number of professionals became actively interested and involved in multicentered clinical trials in order to establish the unequivocal scientific effectiveness of new medical procedures, products or agents. The exploding market of ever new medical devices strengthened the desire of the professional to have some kind of criterion to make a responsible choice.

The fourth and last thrust comes from the national health administrations in cahoots with the scientific communities. As Bunker, Fowles, Schaffarzick and Relman (3) pointed out, large amounts of monies were traditionally spent on medical research, more recently followed by its twin "Development". R & D, however became increasingly isolated from the longer-term effects and relevance of innovations and subsequent applications. With the larger-growing investments needed for a new break-through, such as e.g. the NMR, the necessity for more structural evaluation became apparent. While R & D budgets went skyhigh in the seventies, evaluation monies remained very scarce indeed. Analoguous to the pharmaceutics system, more and more health administrations are considering licensing-systems for large or high-impact new tech-

niques, the licences to be provided on the basis of evaluative
proof of effectiveness.

The heart of the matter of the emerging evaluation-system
is the desire and the need for an objectivation of norms and
standards of medical and health practices. Returning to the sub-
system of norms of input/process/output/outcome in paragraph 6,
the first concern regards the development of criteria that can
be utilized to assess a specific medical activity. So far, the
quality valuation, e.g. in a malpractice suit, remains per se a
subjective and personal matter; objective arguments only enter
the case in a secondary capacity. Systemic evaluation, then, will
not be possible. Therefore, to assess the correct position of a
medical procedure or result, one has to reverse the sequence and
to start from the objectively formulated position which then,
secondly, can legitimately be mitigated down or up according to
the subjective circumstances of the specific case.

Or, to phrase it differently, the objectively arrived at
norm of medical practice (e.g. a sequence of a certain number of
lab-tests specified as to confirm or disprove an indication-
hypothesis) is the standard to which each practitioner or group
of practitioners in principle will have to conform and according
to which their practice will be assessed; in casu certain special
circumstances may then afterwards subjectivize the case and serve
as a valid excuse -deculpability- that in this particular instance
the norm would have been too high. Thus the norm stands and can
be "excused" rather than lowering the norm itself because in
these and those conditions it could not be met.

It will be clear that the development of objectivated standards
of procedures and results into a systemic pillar of the health
services system - very much in line with the algorithmic "protocol"
movement - substantially overlaps the in paragraph 2 described
introduction of efficacy and effectiveness! The 100% compliance
with the norm refers to efficacy, while the deviation of the norm
- justified because of the specific subjective circumstances -
concurs with the effectiveness ratio of 60, 70, 80 etc. %. To
render the evaluation system workable, it therefore will be a
conditio-sine-qua-non not only to establish the objective standard itself, but concurrently to establish guidelines for the
limits of the effectiveness-ratios downwards as well.

Effectiveness and efficacy, of course, are just two of the
criteria of health services evaluation. The now commonly accepted
other ones are adequacy, cost-effectiveness, efficiency, and productivity. Assessment parameters like accessibility of care,
acceptability of services, cohesion of facilities, safety, etc.,
are in principle part of these criteria (Fig. 1A).

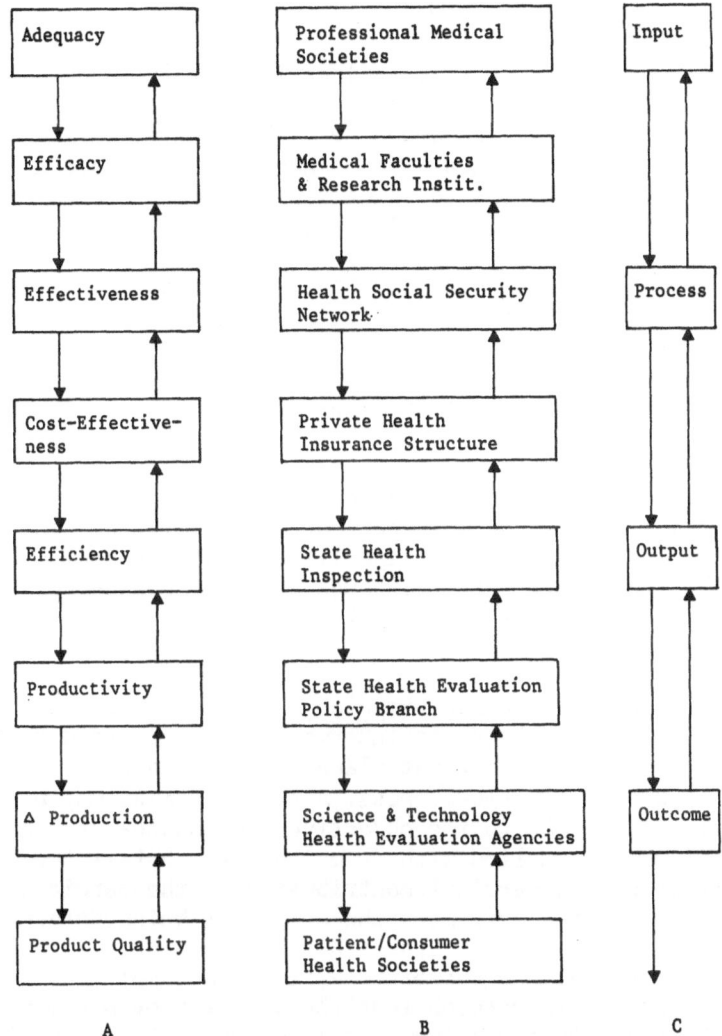

Fig. 1   Sequences of Sub-Systems.

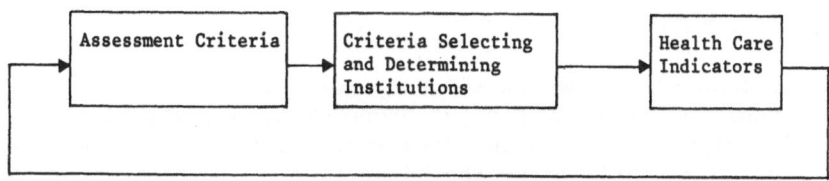

Fig. 2   Health Evaluation Sub-Systems Flow.

3.1. The Assessment of Adequacy

Evaluation as to adequacy coincides largely with the parameters of cost-benefit analysis. All the major drawbacks that
apply to CBA equally hold for adequacy assessment as well.
Adequacy evaluation as opposed to CBA, however, does have a less
significant claim to be able to value the loss or prolongation
of individual life and therefore is less conspicuous emotionally.

The assessment of adequacy is especially needed in view of
the question of causality.

Health services traditionally claim to contribute to the
health of a population. This sounds reasonable, but it may be
less so when the claim is altered in "to be the cause of the
health of a population". As has been made clear by several au-
thors it remains to be seen to what extent what health care is
the cause of what health of what part of the population. In
general it is assumed that something like 25% of our health can
be contributed to health care, although studies showing health
care also to be a cause of disease itself are not uncommon
either (4).

The point, for evaluation purposes, of course, is essential.
The assessment whether a certain intervention can be said to be
valid at all depends in the first place upon its adequacy and
only secondly upon its effectiveness. A triple bypass may be
very effective according to the objectives and standards I have
formulated as to be complied with by my surgery, but if the
intervention is only a marginal contribution to the cardiovascular
health component of the patient - the component being determined
for 95% by variables a triple bypass has no relation to, indeed,
medicine in general has no answer to -, then the health of that
patient would have been more or less the same (or even somewhat
better) without the intervention. The weighing of the causality-
degree of the triple-bypass determinant should, of course, be
specified as to the period concerned, i.e. on a scale of 3 weeks
after the intervention until 4 or 5 years thereafter, as to the R
population/patient concerned, i.e. the elderly (e.g. > 80) or the
young (< 20), etc.. In short, many health services can be superbly
effective, while at the same time being most inadequate.

3.2. The Assessment of Efficacy

Rather than to proceed on the basis of the statistical mode
of the commonly utilized subjective procedures, objectives and
results of peripheral health services systems, efficacy standards
will have to be established by the highest possible centres of
excellence in the field concerned. The systematic labelling,
development and standardizing of which procedures, outputs and

outcomes are possible under ideal, controlled circumstances is a
huge enterprise that will take years to even start with. On the
other hand, at this stage efficacy assessment needs to be introduced mainly as instrument of the "conditioning of the mind" in
order to "conditionalize", to direct and to influence all the
various effectiveness-endeavours, clinical trials and quality
concerns. Efficacy is the starting-point for effectiveness and
efficiency norms, and should fit into the evaluation system
right after the adequacy question.

3.3. The Assessment of Effectiveness

Effectiveness studies, so far, are rare. Cost-effectiveness
studies are becoming popular. The studies, for example, concerning the analysis of intensive care units, of diagnostic equipment
like the CAT scanner, of cancer chemo-therapy, of mass-screening,
of heart surgery, of caries, of megavolt irradiation, etcetera,
generally do not meet the criterion of alternative ways and
means to reach a comparable level of outcome (5). Worse, some of
the studies juxtapose the placebo, rather than what should be
measured against, i.e. the second-best, the third-best, etcetera.

Some exceptions do exist. The studies on renal insufficiencies,
on telephone vs. clinic counselling for hypertensive patients, on
health care programs for the elderly, on day hospitalization, on
psychiatric case finding and on screening for asymptomatic bacteriuria all do utilize the alternative mode of approach (6). All of
them, however, from the very beginning link costs, thereby - in
terms of mere "effectiveness"-polluting the arguments. Effectiveness
analysis in the sense as described supra with regard to cervical
pain, remains a difficult category. Still, it is here that our
evaluative gains will have to be made. Due to the various remuneration systems that link the cost structures of health services
directly to the income-structure of the health professional, cost-
effectiveness analysis will generally carry significant implications
for the income position of the professional or group of professionals concerned. It was this mechanism which frustrated the PSRO-
initiatives in the U.S.; the professional standard review organizations have not been able to disconnect the cost/income link,
thus condemning themselves to failure.

A system which would disregard costs in the first place and
which would primarily be interested in alternative ways to reach
the same or a better effect, however, could well claim to add
to the scientific insight of the professional field under scrutiny
and thereby secure the necessary cooperation of the professional
sub-system. The basis of such analysis would not, of course, be
rooted in individual medical pathology. It would be rooted in a
system of collective patterns of utilization of certain procedures,
their sequences, their mutual interdependence and their results.

The "collective" pathology system often is different from the
individually assessed conceptions of the professional. Medicine
and health care are by virtue still very much individually-
oriented entities. The professionals involved subsequently are
generally unable to reduce their problem to an individual case
as a function of the collective pattern of such cases. It is
this difference that could add to the understanding of the
subject. The pathology of measles and its prevention was and is
well understood; the individual pathogenesis, however, has no
understanding of its collective origin, of the reservoir of
somewhere around 100.000 cases necessary to maintain the in-
fection-chain, of its interdependent relationships with the
epidemiology of malnutrition, with infant diarrhoeal disease,
etcetera. All the latter aspects carry essential implications for
the adequacy and the effectiveness of the individual diagnosis
and treatment of the measle patient.

3.4. The Assessment of Cost-Effectiveness

        Having touched upon some characteristics of the assessment
of adequacy, efficacy and effectiveness, the heart of the health
care evaluation system consists of the theorem of cost-effectiveness. Cost-effectiveness in practice relates to two different
organizational concepts. To the system of hospital administration,
health management, it frequently refers to efficiency; i.e. two
notions, effectiveness on the one hand and efficiency on the other,
cost-effectiveness equalling efficiency. This is the meaning of
CEA in, for example, the CEA-programmes of the Ontario Hospital
Association, where cost savings are the prime objective, to be
realized by revising managing methods for such topics as hospital
housekeeping, utilization rates of operating theatres, illuminating
duplication of materials and reducing unnecessary inventories,
conservation of hospital energy, etc. (7)

        The second, more fundamental, CEA-concept concerns the alter-
native effectiveness system, as described above with the example
of cervical pain. In CEA, the costs of the various alternatives
are considered and cost-effectiveness ratios determined. To
illustrate the principle: to assess the cost difference between
the death of a triple bypass 68 year old patient e.g. 12 months
after the intervention and the death of the same patient after
7 months without the triple bypass compared with the effect the
costs incurred (manpower, time, facilities, monies, etc) could
have sorted if and when applied to alternative surgical interventions
with similar objectives for which the means are scarce as well
(e.g. ESRD-interventions).

        While the first concept is a most necessary managerial tool,
it is the second concept of CEA that will enable evaluation to
grow into a sub-system that is crucial to the health policy

objectives of priority-setting on the basis of the relative effectiveness of certain health and medical services, rather than on the basis of costs and absolute amounts of monies allocated per group of intervention. A system of allocating resources according to the highest effectiveness-ratio as compared with other interventions (to repeat: the health or medical consequences compared and valued not in monetary terms) is, of course, only useful to set priorities among the as such already accepted means of care; it does not infringe upon the question of adequacy. "Highly effective, most inadequate" remains possible, as observed earlier. The same holds true for the next step, i.e. "most efficient, but rather ineffective".

3.5. The Assessment of Efficiency

The last statement has been true all too often for those professionals who are obsessed with demand without questioning the origin of demand. Tonsillectomies, hysterectomies and dental care are conspicuous examples of increasingly efficient utilization of health care resources, but where the effect in terms of the health of the patient will often have been negative (8). Efficiency as a managerial health and medical practice tool is obvious and does not need much clarification. Efficiency in terms of the first cost-effectiveness concept needs very much attention in its own right. Where the promotion of "alternative assessment" may be difficult, the promotion of singular effect measures remains equally embryonic. The activities of the health and the medical professions will need to be structured according to an assessment system which dynamically and permanently questions the necessity, value and causality of resources with respect to the effect that is asked for. Here the pre-formulated effect is not compared to other possible consequences; the singular effect is accepted as such, while now it is the resource side that needs alternation to attain the same singular effect at lower costs. The mentioned OHA programme comes into this category.

A distinction should be made with regard to productivity. The system of quality control circles, as for example common in Japan, surpasses the efficiency concept. It strives for the optimal range of the costs side and the effect side, the nature of the costs and the nature of the effects, however, remaining equal. Generally it concerns increasing the numerical quantity of the effect and decreasing the necessary resources. Productivity of, for example, a family practitioner's practice should be clearly distinguished from the increase of production itself. Due to the erratic origin of need and demand as functions of the pathological indication which is defined and formulated by the supply side only, productivity gain in, for example, group-practice often is accompanied by an equivalent increase in production itself. The result may well be that total costs, despite much better productivity, actually rise and cost- effectiveness deteriorates.

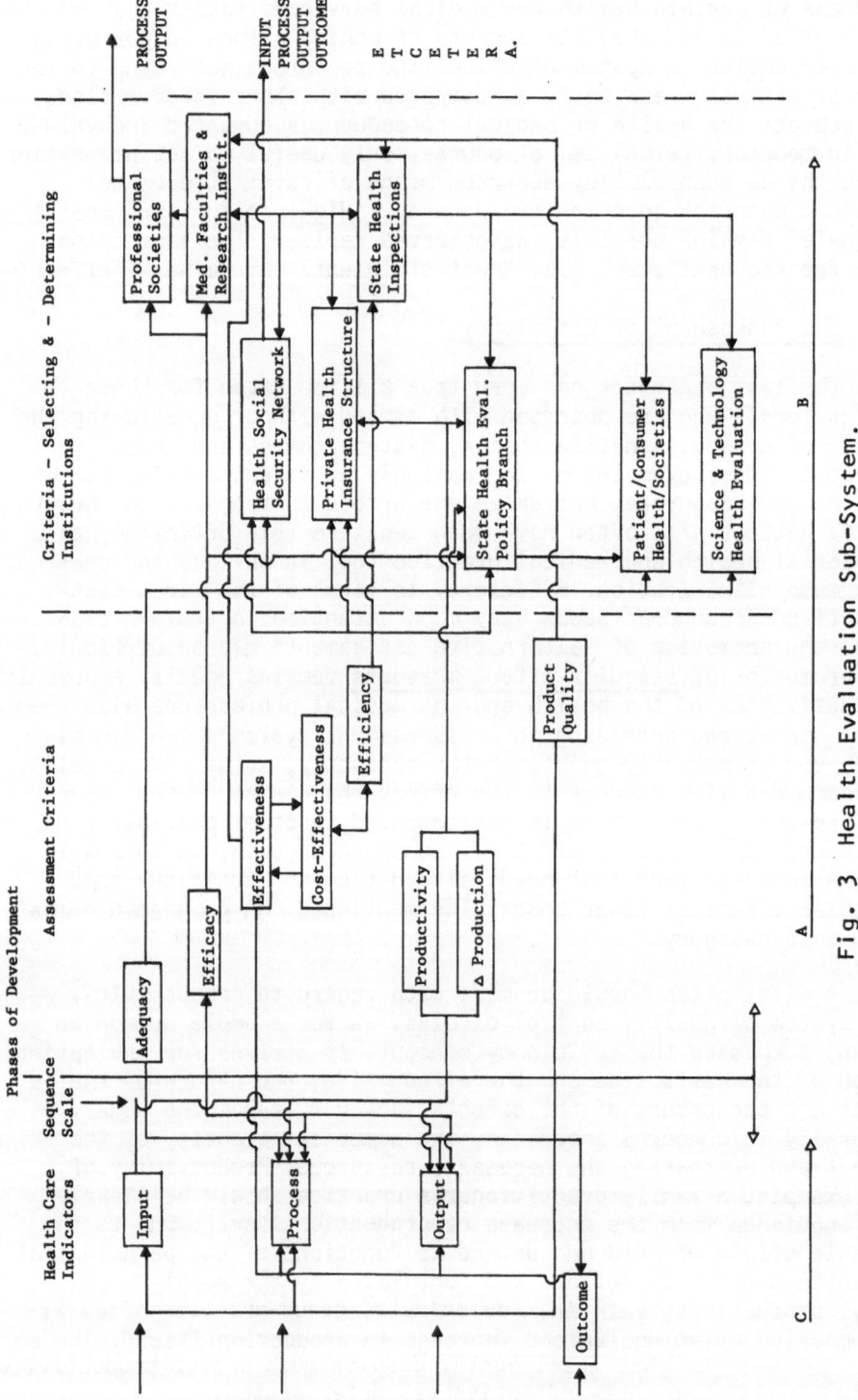

Fig. 3   Health Evaluation Sub-System.

4. Systems of Effectiveness Evaluation

The present systems of evaluation of health services do not emerge according to the classifications and criteria set out above. Evaluations are not scheduled in terms of adequacy, efficacy effectiveness and so on. Assessment systems that have acquired a reasonable state of success have mainly been modelled after the major existing other systems of health care. The most prominent ones in this respect are

1. the medical profession system;
2. the health insurance system;
3. the state quality control annex health inspection system. (Fig. 1B)

4.1. The Medical Profession System

The medical profession system can be considered to consist of two subsystems: the professional societies and associations on the one hand and the medical academic and research institutions on the other. The assessment instrument originating from both sub-systems and controlled by them is peer-review. Peer-review, or inter-professional evaluation, is not specifically linked to any of the criteria. Aspects of all the criteria may play a role in whatever constellation, but then they may equally fail to do so and settle for one particular criterion such as e.g. efficiency. This system of "medical audit" derives its standards wholly from the professional assessment of what is to be expected in terms of effectiveness, efficiency, etcetera. Obviously the system is closely interconnected with the remuneration and insurance system, especially if the fee-for-service mechanism exists. (Fig. 3B). Mixtures where, for example, a spurious synovectomy of the knee is exchanged for the genuine diagnostic arthroscopy are quite widespread in almost all countries.

The definitions of "the state of the art" commonly is set by the centres of excellence in the professional field concerned, whether they be departments of medical faculties or specialized research centres. Often these institutions also control the curricula of training, thereby ensuring the intra-professional hold over e.g. the effectiveness criteria to be utilized by evaluation. (Fig. 3A/B). To establish degrees of proven effectiveness, it is within this system that the wide-spread use of randomized double-blind multicentered clinical trials are performed. Despite the often still somewhat weak methodological basis of these trials and despite their need for huge budgets, the clinical trials are a major step forward towards the more systematic determination and systemic incorporation of health care and medical effectiveness within the health services system.

4.2. The Health Insurance System

In most Western countries this system increasingly emerges
as the most powerful advocate and instigator of a series of
health care evaluations. Again it does not conform to the evaluation criteria systematically, but it puts them to use in
several combinations. So far the development of more structural
applications within this system have been:

1. needs assessment, and
2. utilization assessment.

The so-called "utilization review" started from the efficiency
premise, i.e. from the evaluation whether the instruments, technology or facilities funded by the insurance-system were being
utilized according to their capacity. This led to the second
step, i.e. asking whether the technologies - now being expanded
to the procedures and techniques involved - were being utilized
in the proper way; in other words, whether they were utilized
for the ends they were designed for. Whether these ends or the
objectives are valid propositions in themselves, i.e. the effectiveness question in its true sense, now starts to be incorporated
in the utilization-review evaluations. In absurdum the query
would, of course, investigate whether the patient who was operated
upon for the removal of a gall-stone, had a gall-stone in the
first place. This, generally, is presupposed. The upshot of the
absurdity, however, led to a system of review to evaluate the
frequency of utilization, a frequency which then started to
display enormuous differences between the one hospital and the
other, the one specialist and the other, the one city or region
and the other. Naturally the insurance system, paying for the
high-frequency variations as well, became keenly interested in the
evaluation of the necessity for these differences.

Although in principle still confined to cost-effectiveness
analysis, this interest implied much larger scopes of assessment.
The interest represents the first difficult step towards the
system of need-assessment. Need-assessment parallels evaluation
on the basis of adequacy. It interferes with the effectiveness-
position of the medical diagnosis and indication from the former
peer-review system (Fig. 3A/B). It demands the established
causality not only between the performed health or medical procedure or technology and its consequences (Fig. 3C & Fig. 4), but
more fundamentally between the proposed interventions and the
presented pathologies.

In theory the evaluation starts with the epidemiological
analysis of regional morbidity and mortality patterns and the
health services needed to improve on these patterns. In practice,
however, this has not been possible to organize, if only because
of the empirical nature of the health services already functioning

in the region. More practical, as has been shown by the CON-reviews
(CON = Certificate of Need) of the American health insurance
system, would be to start from the differentiations at micro-level,
e.g. hospitals, group practices, departments, HMOs, etcetera.

The need-assessment then could become a logical sequence to
the utilization-review on the basis of effectiveness variations.
The argument a-contrario would open a macro-system where a decrease
or an increase in a group of certain specified health services
would have to be accompanied by an inversely proportionate change
of the morbidity- or mortality-patterns concerned, depending upon
effectiveness-elasticity of the needs under investigation. Within
the insurance system this kind of assessment could be most
valuable, since health services, on the basis of adequacy and the
effectiveness argument, could be reduced down to the point where
it collectively would show in the morbidity-patterns concerned.
(Fig. 2 with fig. 3B-C). This health- or disease-status-index (HSI)
approach meets, however, a number of obstacles from the quality
control and inspectorate system.

4.3. The State Quality Control and Inspection System

The state inspectorate system is by far the oldest quality-
of-care assessment mechanism. Formally most inspectorate systems,
apart from their disciplinary and prosecuting functions on the
basis of open negligence, personal bodily damage or specific complaints, were also to evaluate medical and health activities
actively and in their own right. Materially this mandate rarely
materialized. The assessments undertaken normally contain a
criteria-mix of adequacy and effectiveness. Costs and efficiency
seldom enter a case. The legal standards of the medico-legal
norm-system play a prominent part in the inspectorate quality
control, with regard to e.g. proper behaviour, privacy, record-
keeping or commercial competition with fellow-practitioners.

Increasingly, however, these standards become of a secondary
nature and are being considered within the context of the quality
assessment criteria. Record-keeping has grown to be a variable
in the correct algorithm of certain diagnostic and therapeutic
procedures. Negligence at a rural delivery by a family practitioner
is being tested against the effectiveness-norm of a-minimum-of-
30-deliveries-a-year to be performed in order to maintain pro-
ficiency. Commercial lab-operations now are considered on the basis
of costs-effectiveness qualities as well. State control agencies
are ever more pressed to actively pursue and trace circumstances
and practices which endanger the adequacy, effectiveness and
efficiency of health services. Iatrogenic and nosocomial afflictions
belong to this category, as do, for example, lenght of stay,
minimum or maximum patient-load (e.g. for one anaesthesist during
surgery), total requests for lab-tests, ratios of therapeutic vs

palliative services for e.g. chronic disease, minimum and maximum
ratios of the various manpower proportions, e.g. the number of
nurses, internists, lab-technicians, etc. in a hospital or the
mutual proportion of e.g. fysiotherapists, family practitioners
and specialists in a certain geographical area, and so on.

Even the quality control state agencies that are willing and
interested to pursue these ends, generally lack the capacity and
the expertise to do so. In many countries, however, the in paragraph 3 described thrusts for evaluation priorities and techniques
are a major source of support for the agencies in this respect.
Indeed, it can be observed that the evaluative effectiveness
system is on its way to become an important health policy instrument more powerful than the original inspectorate system of
quality control and that the state agencies, with regard to the
criteria for assessment, are jumping on its bandwagon. A more
close cooperation and possibly integration of the various new
health-evaluative undertakings and institutions with the traditional quality control state and regional agencies could well be
envisaged. The subsequent assessment system would certainly not
be second to any of the other two systems described.

5. The Overall Assessment System

The various activities, approaches and efforts with respect
to the assessment of health care can be said to form an organized
set of doctrines, ideas or principles intended to explain the
arrangements or working of a whole and at the same time "to pertain
to the (health services) system as a whole", i.e. to form a system.
As may be equally clear meanwhile, the various components of the
system sometimes are organized and arranged erratically and often
bear a complex mutual relationship. Within the overall assessment
and in the mutual exertion of influence among the sub-systems, the
most binding principle is the theorem of (cost-) effectiveness.

Grouping the various components yields the following organized
set of mutually interdependent and interactive sub-systems (Fig. 1
& 2):

A. Assessment-criteria Sub-system (Fig. 1A & 2)

1. Adequacy
2. Efficacy
3. Effectiveness
4. Cost Effectiveness
5. Efficiency
6. Productivity
7. Δ Production
8. Product-quality

Product-quality assessment refers to the evaluation of the safety of certain health and medical devices, instruments and technologies. It is also concerned with the commercially competitive properties of the technologies, i.e. which is the "best" for you in market-terms. Although this does have relevance for efficiency, the assessment explicity stays clear of effectiveness considerations and as such is less pertinent to the subject.

B. Sub-system of Criteria-Selecting & -Determining Institutions (Fig. 1B & 2).

1. Professional Medical Societies
2. Medical Faculties & Research Institutions
3. Health Social Security Network
4. Private Health Insurance Structure
5. State Health Inspectorate
6. State Health Evaluation Policy Branch
7. Science & Technology Health Evaluation Agencies
8. New : Patient/Consumer Health Societies

C. Health Care Indicators Sub-System (Fig. 1C, 2 & 4)

Having identified according to which criteria health services are to be evaluated and which institutions structurally are concerned in this respect, the object of evaluation, i.e. health care, will have to be measured and as such to be incorporated into the system as well.

The measurement is best to be structured in terms of indicators. Since a good classification is the well-known S.A. process distinction of input - output - outcome indicators, the following schematization will suffice for summary (Fig. 4).

1. Input-indicators
2. Process-indicators
3. Output-indicators
4. Outcome-indicators

Instances of indicators would be, for example:

1. Input   : - available beds
              - usable hours of manpower
              - allocated budgets

2. Process : - occupied beds
              - o.p.d. attendance
              - lab-tests
              - monies spent

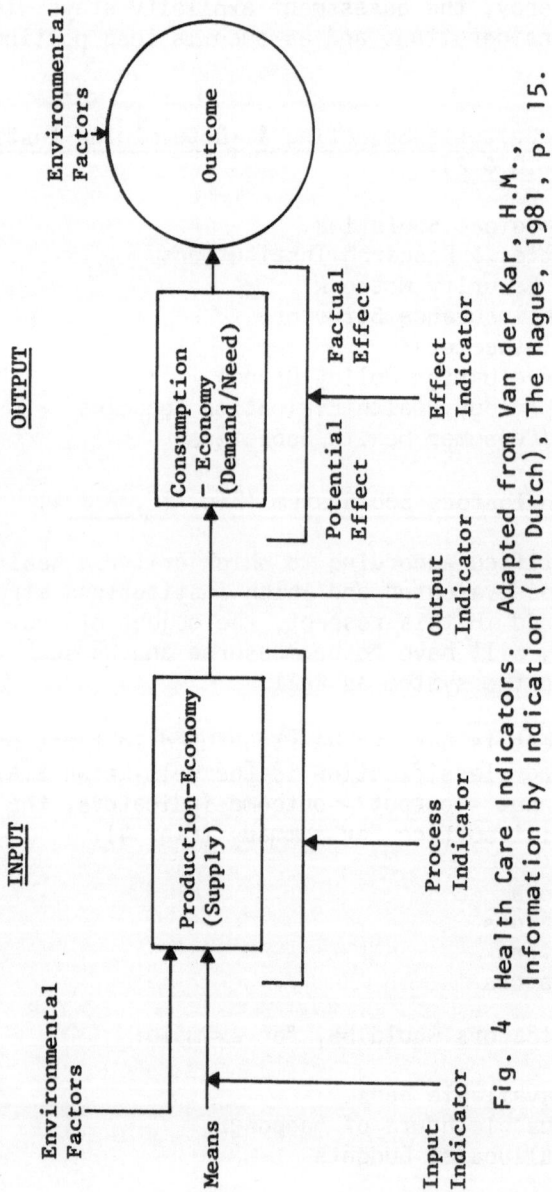

Fig. 4   Health Care Indicators. Adapted from Van der Kar, H.M.,
Information by Indication (In Dutch), The Hague, 1981, p. 15.

3. Output   : - patients discharged
               - vaccinations performed
               - referrals made

4. Outcome   : - patients healed
               - patients died
               - preventions caused
               - infections diminished

The effect-indicators, in accordance with the distinctions of the assessment-criteria sub-system (A), either follow the output-indicators (e.g. the synovectomy example) or the outcome-indicators (e.g. the example of the cervical tumor therapy, i.e. longevity and 5-year survival rate).

The sub-systems A, B and C, then, can be displayed in mutual relationships, clarifying their reciprocal positions with respect to the effectiveness determinant of health care (Fig. 1, 2 and 3).

6. Conclusion

The underlying view is meant to spur the discussions on (cost-) effectiveness as a tool of health services evaluation and how the present conglomerate of unstructured ad-hoc elements that are involved can be structured

1. into a coherent and consistent sub-system of effectiveness evaluation, and

2. into a viable systemic part of the overall health services system.

While recent experience shows that the application of systems approaches to health services can successfully be of paramount importance with regard to reorienting the health & medical care process at the micro-level (e.g. CEA or efficiency criteria with regard to diagnostic or lab algorithms or to hospital patient-flow stochastic models), time and again it proves to be much harder to provide applicable systems' concepts for the planning and management of health services systems and sub-systems at the macro-level.

If health services evaluation, with effectiveness-derived criteria at its heart, is ever to come of age, the "conditioning of the minds" of quality-of-care professionals and health policy makers in terms of dynamic S.A.-concepts applied to macro health services models (including a model of a health services evaluation system) as shown in figures 1 - 4, will be an absolute prerequisite. The discussions, papers, clarifications and new approaches that were presented at the ARI-HSS Conference, provide valuable stimuli to this end.

References

1. Office of Technology Assessment, The Implications of
   CEA of Medical Technology, Washington, 1980, pp. 3-5.
2. A.R. Prest, and R. Turvey, "Cost-Benefit Analysis - A
   Survey", The Economic Journal, 75, December 1975.
3. A.S. Relman, "An Institute for Health Care Evaluation",
   J.R. Bunker, et al., "Evaluation of Medical
   Technology Strategies", New Eng. J. of Med., Vol. 306,
   No. 10, March 11, 1982, pp. 610-624.
4. K.L. White, T.F. Williams and B.G. Greenberg, "The Ecology
   of Medical Care", New Eng. J. Med., 1961, Vol. 265,
   pp. 885-992.
   I. Illich, "Medical Nemesis - The Expropriation of Health",
   London, 1975.
5. G.E. Thibault, A.G. Mulley, et al., "Medical Intensive
   Care: Indications, Interventions, and Outcomes",
   New Eng. J. Med., 1980, Vol. 302, No. 17, pp. 938-948.
   J.R. Bartlett, et al., "Evaluating cost-effectiveness of
   diagnostic equipment: the brain scanner case", Br. Med.
   J., 2 (1978), pp. 815-820.
   R.J.J. Berry, "Dare we count the cost of cancer chemo-
   therapy?", Lancet, 2 (1978), pp. 516-518.
   C.B. Clayman, "Mass screening: is it cost-effective?",
   J.Am.Med.Ass., 243 (1980) 20, pp. 2067-2068.
   S.A. Finkler, "Cost-effectiveness of regionalization:
   the heart surgery example", Inquiry, 16 (1973) 3,
   pp. 264-270.
   H.S. Horowitz, and S.B. Heifetz, "Methods for assessing
   the cost-effectiveness of caries preventive agents
   and procedures", Int. Dental.J., 29 (1979) 2, pp. 106-
   117.
   C.R.H. Penn, "Megavoltage irradiation in a district
   general hospital remote from a main radiotherapy unit -
   a cost effectiveness study", Health Trends, 13 (1981),
   pp. 26-28.
6. E.M. Bertera and R.L. Bertera, "The cost-effectiveness
   of telephone vs clinic counseling for hypertensive
   patients, a pilot study", Am. J. Publ. Hlth, 71 (1981) 6,
   pp. 626-629.
   N. Doherty and B. Hicks, "Cost-effectiveness analysis and
   alternative health care programs for the elderly",
   Hlth. Serv. Res., 12 (1977), pp. 190-203.
   W. Guilette, et al., "Day hospitalization as a cost-
   effective alternative to inpatient care: a pilot study",
   Hosp. & Comm. Psychiat., 29 (1978), pp. 525-527.
   M.K. Isaac and R.L. Kapur, "A cost-effectiveness analysis
   of three different methods of psychiatric case finding
   in the general population", Brit. J. Psychiat., 137
   (1980), pp. 540-546.

G. Rich, N.J. Glass and J.B. Selkon, "Cost-effectiveness of two methods of screening for asymptomatic bacteriuria", Br. J. Prev. & Soc. Med., 30 (1976),pp. 54-60.

Th.C. Eickhoff, "General Comments on the Study on the Efficacy of Nosocomial Infection Control", SENIC Project, Washington, 1979.

F.Th. de Charro, "An Evaluation of Pro's and Con's of Alternative Strategies for End Stage Renal Disease", (in dutch), University of Rotterdam, 1981.

7. J.R. Haslehurst, "The Ontario Hospital Association's Cost-Effectiveness Programme", World Hospitals, Vol. 17, No. 4, 1981, pp. 31-33.

8. A.F. Lewis and W.L.J. Modle, "An Approach to Efficacy in Health Care", Health Trends, 1982, Vol. 14, pp. 3-8.

D.L. Crombie, "Proceedings of the Royal Society of Medicine", 1977, Vol. 70, pp. 407-410.

HEALTH SERVICES SYSTEMS AND INFORMATION SERVICES

A.S. Härö

The National Board of Health

Helsinki, Finland

0. Introductory remarks

Prologue : To discuss health systems and information services
at the level of theories is definitely a useful exercise. It is
a truism that "nothing is more practical than a good theory" (as
B. Russel once said). But not all are interested in theories and
in order to deserve the attribute "good" they must be easily applicable to pragmatic real life problems. Another approach is to describe analytically what has happened in one single nation, well-
known to the speaker. A natural prerequisite is that the circumstances are also known well enough to all other participants. The
actual fact is very few national services are well analyzed and
presented as "systems". Those who live inside of the system, or
actually belong to it, have usually a biased view of circumstances.
There are values and priorities which are for "insiders" so self
evident that there is no reason even to mention such aspects. But
the same is true when "outsiders" are describing the service system
of other nations. In any case much space is needed for system
descriptions.

The focus of interest in this paper is the information -
actually in which way the information services are helping to
make from separate uncoordinated units a "whole" which deserves
to be called a "system". Theoretically speaking nearly every
national arrangement satisfies the minimum requirements of a
system. But very few, if any, pass a really strict analysis using
as criteria the overall achievements towards goals which are
relevant at present and especially with the future in mind. To
mobilize the whole "health system" toward approved long-term

goals like "health for all by the year 2000" is a difficult task
for all nations. Actually all countries represented here have
formally adopted this goal. As members of WHO they have promised
to give a high priority to it in their future health policy. A
comprehensive view of "health" and articulation towards equity
are the main elements in this policy.

1. Information services as a tool:coordination and management

There are societies in which the decision processes are very
centralized but some others in which the ownership and managerial
responsibilities are delegated into the periphery as far as
possible. But there are tools which serve as a guarantee that the
"whole" is functioning as a purposeful system and that the central
authorities can influence enough the purposes. Information "rights"
are a useful element both directly and indirectly in this connexion.

The element that makes purposeful interaction possible is
information. It is natural to introduce systems concepts into
the management and administration of an organization, institution
or company, etc. The expression "management system" is at present
a normally used one. Its objective is very clear - to guide the
system as a whole in the expected direction or to defined goals.
The term "management by systems" (or systems management) has a
somewhat different content. The purpose is to stress that the
activities are understood as systems of functional entities instead of as formal organizational structures. The word modern which
is often used in this connection refers to the managerial styles
in which the objective of the organization are quantified and all
possible means are used in order to achieve these.
The organization as a whole can have as the main objective the controlling of its area of interest. In any case internal control is
the main function of the unit which has assumed the managerial or
administrative responsibilities. The unit should be able:
- to detect problems and difficulties
- to make meaningful decisions
- to control that the decisions will be materialized
- to take care that relevant information and knowledge are
  available concerning the system and its environment.

The basic and unavoidable function of management is making
of decision. Decision-making is a process which converts information to instructions for the system in order to improve
its performance. Information has a central role in this process.

General guidance and control is not feasible without statistics, analytic studies, expert opinions etc. Without proper information most goal-oriented managerial functions are impossible and
the intentionally set objectives of the activities will not be
achieved. It is hardly necessary to mention that an elementary
part of the decision-making process is planning.

A formal planning process without solid facts and proper information is hardly possible at all.

2. Some definitions

Health, system and information are abstract concepts and very
different services can be labelled with these as being "health information systems". In this paper the following definitions are
applied:

Information system:

A generic name for all the activities or elements which together take care of the collection of data, their processing to
information, presentation etc. Information system can include,
except statistical services, in addition e.g. research activities,
experts, library services and a subsystem for information processing.

Information services:

The main task of information system is to serve the decision-
making process in the organization. The word service in this connection stresses this aspect.

3. National health information systems

The word national is in this connexion used having in mind
the needs of a nation or the society as a whole. In principle the
health information system mainly reflects the boundaries of health
systems as understood by those who are responsible for managerial
activities. The more inclusive the concept of health system, the
more comprehensive is the information system. Actually the service
systems can be tested using information services as indicators.
One can not be made responsible for managerial activities without
being able, at least in theory, to be properly informed. If health
authorities have no information of something and even no powers
or resources to be informed, they can not be made responsible for
mismanagement. Sometimes it is understandable that the higher
level authorities are not interested to "see" the problem which
might give them the responsibility to act. Sometimes the lack of
interest might reflect the fact that "who-ever defines the problem
selects the solution as well". There is reason to suspect that
some health related problems are intentionally kept outside the
health system in order to apply solutions more suited to some professional groups.

The organizational structure has a very great influence on the
information needs. Often legislative clauses make some information

subsystems obligatory. Information system is said to be a "mirror image" of the organizational structure and organizational goals.

The comprehensive planning of an information system for an organization is closely related to replanning and restructuring the organization as a whole.

Ackoff has presented criteria for interrelationships between the organizational systems for decision-making, planning and information:

- the optimal structure of an organization minimizes the decision-making and planning system;
- the optimal decision-making and planning system minimizes the information system.

An unbalanced and impractical information system accordingly indicates that there might be shortcomings in the basic structure of the organization as a whole. A complicated system requires complicated information systems. The weakness in an information system can primarily reflect weakness in the decision process.

All health information systems have numerous wellknown weaknesses and limitations. Usually the information services are not seen as a "whole" and each of its components are discussed in isolation. Very naturally, such components also have a tendency to function as independently as possible - there are e.g. research units which stress the independence or liberty of service more than any other value related to their functioning. In these type of questions the only normative element should be the problem to be solved. There are questions in which a complete independence is obviously fruitful. But there are numerous others where a planned cooperation and division of labour is much more useful.

4. Criteria for good information service.

A purposeful information service is a natural prerequisite for an effectively managed health system. One actual problem is often the criteria of a good service. At least the following problems need consideration in this connexion:

          Problems of : - organizational location
                        - coordination and cooperation
                        - information content
                        - methodologies and procedures
                        - data processing, analysis
                        - communication and presentation.

Each of these will shortly be discussed with reference to experiences in Finland. The problems in the list mentioned above, actually form a framework for purposeful activities to develop information services in a planned manner.

In principle there are two strategical alternatives to most planning problems. We can either do our utmost in order to achieve acceptable objectives with existing managerial tools. Another approach would be to reorganize and strengthen such components in the system which can achieve the targets. In the first approach the focus of interest is directed toward the "end" and in the second towards the "means". Both of them have their merits and limitations, and the circumstances define what is optimal. If the highest authorities are really interested and willing to achieve long-term goals, an investment to means is very logical. If the progress is mainly the responsibility of lower level administrative staff, it is usually difficult to introduce any major changes in resources or structure. The two extreme situations require different strategies and to some extent this is also reflected in the solutions of all the other problems.

5. Information strategy problems

Some of the strategic problems have already been discussed. But there are many others like the relationship of information to policy priorities. Information services should be especially strong in areas where the organization will be active or where problems are to be expected. Limited resources can e.g. either be focussed to a few main problems throughout the nation, or concentrated to a few densely populated places. Health policy can be either active or passive and this definitely influences the whole information climate from top to bottom.
The role of information service planners is to analyse the policy and formulate a clear image on which clearcuts of knowledge it is based or at least should be based. And some of the most relevant alternatives and perceived problems which might grow acute in the future must also be considered. The most suitable managerial level which can have competences in such policy analyses naturally varies, but experiences in informatica are definitely not the best background for this level. A thorough understanding and shared views are especially important because the construction of information sub-systems usually takes much time. If such strategies are not considered in due time the decision must later-on be made mainly on the basis of guesses.

Because resources are in most cases limited, the essence of the problem are the priorities, in which order and/or in which time. A strategy which tries gradually to develop the existing sub-systems to a functioning "whole" is generally much more realistic than to build a complete new system from the very beginning.

6. Organization:locations, structure, etc.

Everyone experienced in practical management knows that location in the hierarchy, formal responsibilities, etc., markedly influence operational achievements. In this case, we have in principle at least three different options to be considered:

- We can locate the responsible health information unit rather high in the hierarchy or do the opposite which means a subordinated bureau located low in hierarchy.

- Information services can also be integrated with some other service unit, like a planning department, or more or less completely disintegrated in such a way that each service unit has its own, specially oriented, sub-system with minimal links to other sub-systems serving health management.

- The third aspect to be considered is to what extent the information system is located inside of the organization, or relies more on "outsiders". The outsiders can be e.g., university departments, ad hoc research teams but also general statistical services can sometimes function as "outsiders".

There cannot be any generally valid recommendation how to solve organizational problems at national level, but definitely much attention should be devoted to these kind of problems. There is reason to stress that an information system is composed of functionally linked units. There is no need that all these form an administrative unit, like departments, but there must be some coordinator who takes care of the links and inter-connexions. There are many ways to solve this problem, and of course the solution must be suited to the general organizational "climate" prevailing in the country.

A functioning information system cannot be managed without good communications and a reasonable degree of authority. The main thing is that there exist such linkages between components and elements that it is justified to label the "whole" as a system.

7. Problems of coordination and cooperation

One basic prerequisite for coordination is naturally the logical organizational configuration discussed in the previous chapter. But there are some other aspects which need attention. Some of these are proper definitions, classifications, areal specifications etc. Very often the reason for gaps in the available information is the basic legislation that, for some reason, focuses attention to details which might have had some relevance

in the past, but are very secondary from the point of view of
purposeful management at the present stage. Legislation has a tendency
to stay relatively long periods, but real life changes continuously.
Too detailed clauses in the law can effectively prevent the practical cooperation of information services with the active managerial
staff. The energy of information experts goes to the secondary aspects and there are not resources for the real, acute questions.

Another problem is the fact that various groups of information
specialists have a very different image of the information system
in their mind. To some the system is composed mainly of traditional
statistics. University researchers are oriented to consider scientific research as an activity which is not intended to serve at all
any managerial purposes. Also it often seems difficult to produce
relevant economic information which serves the financial policies
of management and especially planning activities. Accounting is an
activity with its own traditions and these are not always suited
to a coordinated functioning in the framework of national information
services.

All these and numerous other coordination problems are obstacles
which must be solved in order to develop a functioning service.
These types of problems cannot be solved without involvement of top
level management. Actually this is one of their main duties.

8. Problems of information content

Of the many qualities of a good manager, two vital ones must
be emphasized : an understanding of the overall objectives of the
organization; the ability to use information. The first one is
essential if there is anything to manage; it is also the focus of
information provision and utilization. The elementary function of
management is to translate the information to instructions. How
could this be done without properly organized information services
which produce relevant information? Accordingly the manager is the
main user of information or at least the main representative of the
group of users. The information content, the relevancy, is the aspect in which the managers should have the essential interest.

One commonly found relic from the past is the attitude to see
information services very narrowly and only a beginning area for
statisticians. Some numbers are needed but "let the statisticians
plan something useable for us" describes well the way of thinking.
There might be exceptions but a specialist in statistical methods
and/or data processing technology cannot be the right adviser in
the problem of what should be known. But on the other hand, the sta-
tistician is the right expert for deciding how the data should be
collected and processed. It is primarily the task of decision-makers
of managers to take care that both these aspects - the content of

information and the applied methods - are taken into account in a
balanced way in the process of planning information services.

Information needs cannot be evaluated by someone who has no
possibility to go deep enough, e.g. an expert in statistical methodology only.

It is said that "to know what should be known" is an easily
solved problem. The opposite is more true and the whole competence
of senior managers is needed to select such information which is
critical, having in mind the decisions to be made, or intentionally
not made. Critical is also information which might be used only in
difficult circumstances, and situations which require "crisis management". Enough attention should be devoted to the environment of
the system, to trends which might become dangerous or are indicating
new opportunities.
When seen from the point of view of the decision-making process,
many of the traditional attributes of information loose in importance
and some others are often more central. Rough, but in principle correct estimates of circumstances are much more important than exact
measurements too late to be usable. In daily operational management
the information should be available more or less immediately. In
strategic problems and in formulating the health policy issues,
solid trends, relevant analysis and thorough understanding are much
more vital than immediate availability of data. Of course the objective should be reliability and timeliness, as well as other values
like validity, but in real life a compromise must be made so that
the information serves the process, not the opposite.

9. Systems analysis and modeling

Planning the content of any information is a difficult task
and involves both users of information and technical experts. To
analyze and describe in detail decisions as they are actually made
in real life situations, is obviously impossible, but much can be
achieved if the most important items can be selected and presented.
This aim can be reached in different ways but one recommended procedure is to analyze and simulate the programmes, activities, policies, functioning of the service units, or anything in which decisions in one or other form can be considered to be possible. Actually this kind of system analysis activity closely resembles the
exercises in schools of business or in military schools where special "games" are played in order to train. In reality the outcome is
some kind of very crude model which has numerous meaningful roles.
If carefully made it explains the functioning of the organization
as a "system" and classifies its interrelationship to the environment.

A health system includes elements which belong to the environment of formal health services systems. In a modelling exercise at least some of the implicite values must be stated openly.

The word model has many meanings but in this connexion - in order to know what should be known - this simply means a systematic way of indicating what activities belong together, which can be controlled and which measurements indicate the success or failure. It can be simply a list of relevant items or an elaborated organogram. The only vitally important aspect is that the result helps to give a consensus concerning the information which is relevant, the necessary accuracy of data, and the risks inherent in using estimates instead of measurements etc. And it is hardly necessary to mention that there are numerous "constants", figures and measurements based on trivial and generally understood relationships, which in any case should be known. Good examples are e.g. such data which are based on clauses in the legislation.

Generally speaking the area to be modelled is "soft" and contains more or less abstract elements. There are only a few more solid or fixed points which indicate the structural elements of the model(s). One of these is a list of existing or postulated problems. Another solid starting point is to locate "decision makers". In this connection this group should be much broader than the formal undersigners and cover all possible interest groups which can influence or should influence the decisions to be made or intentionally not made. The expression of "problem owners" defines very well these groups.

The focus of interest is not the formulation of decisions to be agreed upon, but only the information items or indicators needed if anyone forms opinions concerning the problem. For this limited purpose relatively crude models are serviceable. But even such models can be constructed only by persons who thoroughly know the actual problem. It must be stressed that decision-makers, or more generally speaking "problem owners" being users, are the leading group in this activity. Without their help the other possible participants do not know which information is critical and which is not needed.

Experts from outside can be helpful but the planning of information services should primarily be carried out inside the "system". Of course due attention should be given to environmental aspects and alternative approaches and goals. There is hardly reason to expect that any organization can evaluate its ultimate end without some bias. This limitation must be noticed and confessed. The achievements of ultimate goals is a task which is more suited to "outsiders" - not to the inside information services. This type of systems analysis, modelling, decision, simulation activity

is applicable at any level of government(local, regional or central),
but it is necessary to have in mind the compatibility, comparability
and capacity to aggregate at each level.

To activate present generations of managers to planning activities described in this chapter seems to be difficult. There are
areas which are relatively well covered by information services and
on the other hand the information aspects at numerous health related
problems have never been fully explored.

10. Methodological and technical aspects

The previous chapter assumes that the prerequisite for a good
service is cooperation, or dialogue, between the user and the provider. Before the technical experts can use their skills to assist
the policy-making process, the manager must be articulate about the
operational urgencies, the objectives to be achieved and which are
the strategies and main policy issues. The basic requirement concerning the manager is that he/she is able to translate all these
into terms which really mean something to technical experts. Together they must convert these statements into exactly formulated
information requirements.

To know what is going on in exact terms is sometimes vital,
sometimes trivial, but in any case the decision process always has
in addition elements which are non-tangible or at least non-quantifiable. Actually, it seems to be true that in most complicated problems the non-quantifiable "soft" information has a decisive role.

The basic roles are not always honoured and applied in practice.
In some cases there does not exist any clear picture of the situation and objectives to be followed are not given. Often the decision-
makers are not willing to describe what strategies are under consideration. The decision-makers and managers on the one hand and the statisticians and the methodological experts on the other hand have
quite different training and orientation. Another reason, making the
dialogue difficult is related with the location in hierarchical
structure. It is difficult to have informal connexions which would
fully compensate a more formally organized dialogue.

One natural requirement is that the information services should,
on the one hand, serve all hierarchical levels (for instance: local,
regional and national ones) and on the other hand function without
undue duplications. Often the same, or comparable, information is
collected separately for different users. To a great extent such can
be avoided if proper attention is given to the problem.

Something should be said about ad hoc research as a part of information systems. The area of the activities which are labelled

research is very wide. But two points need attention in this connexion. The first one is that both the ad hoc research and the statistical subsystems serve the needs of decision-making processes to
discuss and to plan these activities quite seperately is not meaningful. Research is also a scarce resource, requiring special competence
and usually much time and effort. It should accordingly be used for
purposes which cannot be solved with other, more economic, ways. A
well planned statistical sub-system which contains e.g., data banks
serving managerially critical sectors, can produce useful information much more easily than is usually thought. Considerable savings,
and in addition better services will be achieved if the two approaches are planned to compensate the limitations of each other. A
great amount of useful research can be based upon routinely collected data, especially if available in data banks at all, or registers.
In many cases the routine data are not analyzed at all, or with only
one special aspect in mind. A differently oriented "scientific" analysis could give much additional, or in other respects useful information.

Much controversial opinion exists concerning the problems related to data aggregation. Linkage, data banks, registers and related
concepts have some special features but in general they reflect
different approaches or strategies of how to produce statistical types
of information. The traditional way has been to use forms in which
e.g., visitors, patients, cases, results of tests, etc., are summarized (or aggregated) for a certain period of time. The standard routine is to have periodical reports in which the basic data are in an
aggregated form. As such this might serve some information needs
very well but there is one unsolvable problem. The introduced form
dictates that the basic data can not be regrouped in order to produce more "tailor made" information.

It is also possible to select the opposite strategy. This means
that basic data are kept as seperate entities and they are aggregated
only when required. This type of statistics, usually called registers
(data archives, data banks) is naturally much more flexible and can
produce information which is tailor made for the problem. Larger registers require in practice many resources and some form of modern
dataprocessing is usually a prerequisite. Registers should be inita-
ted only after very careful considerations in order to support a
defined critical issue. Much resources can easily be wasted and it
seems that it is sometimes very difficult to stop such a function
when once initiated.

The registers have another quality which adds a new dimension
to information services. The data of different registers can be
linked together with the help of systematically used identifiers.
There exist medical problems which hardly can be solved without
linkage of data at personel level. But most managerial needs of

information can be fulfilled in more practical ways of linkage. A
systematic use of areal, institutional, occupational etc. codes makes
it possible to link in a purposeful way.

The plea of most administrators and policy makers is to have a
few indicators like the GNP, that could reflect the changes in the
health situation. The idea is natural but, at least until now, difficult to materialize. Some kind of "surrogates" have been used,
e.g. infant mortality rates and mean expectation of length of life.
It is generally speaking easier to compose indicators which describe
the environment in which the health system is functioning.
Urgently needed, but much more difficult to have, are indicators
which in short time reflect the impact of the interventions. There
are numerous ongoing international activities in this field. This is
quite understandable because, if anywhere so, for international
comparisons reliable health indicators are urgently needed. The basic problem is related to the aggregation of data to one or a few
indexes. The fewer indicators there are, the more details will be
lost. Maybe more influential is the fact that different series of
statistical data cannot be summarized without some kind of weighting. And correct weights seem to be so heavily biased by subjective or specific points of view that no agreement on objective,
"scientific" general weighting formulas have been achieved. One
natural solution is to increase the number of separate indicators,
but in this case the basic idea of one or few series of numbers has
been abandoned.

11. Problems of data processing and analysis

Of the special skills of the technical experts the use of computers is elementary. In the not too remote past the computer was
often presented as the solution to all information problems, as being
"the information system". This was of course done especially by the
salesmen but the idea was successfully marketed both to decesion-
makers, statisticians, and other expert groups. Undeniable many rou-
tines are nearly impossible without computers and definitely it is a
most useful servant in all data processing. The personal computers
or terminals are obviously soon a normal solution also in the health
field. The bottleneck is no more in the computers but in the input
processes. A developmental strategy which is based only on mass-col-
lection of interesting data which the computer should digest into
meaningful information, often fails. The fault has not primarily
been in the computers, nor in the data processing experts, but in
the lack of dialogue and cooperative planning which must precede
critical decesions.

One of the prerequisites of an information service is reliability. As such this is a quite understandable requirement but reliability again has many dimensions. Sometimes it is evaluated very formally,

e.g., on the basis of counting errors or comparable details. Much more dangerous are weaknesses in validity e.g. when basic definitions are not fully applicable or have never been documented or agreed. There are other types of unreliability based e.g. on differences in coverage, use of less correct denominators in areal comparisons, etc. Errors are not to be defended but on the other hand the criticism is often one-sided and stereotype.

Formal exactness is not always the quality that serves best the interest of decision processes. Rather "unreliable" but relevant information can effectively limit the uncertainties if the size and direction of the errors can be estimated.

Often the direction of development is the most important aspect and rather poor basic data can provide useful statistical series that show the trends with reasonable exactness. A good example is the notified cases of sexually transmitted diseases like gonorrhoea. All cases are never notified but still the trends can be meaningful.

The unused information is a common complaint and sometimes quite correctly. Very often, however, behind this is a misunderstanding or at least a one-side view. Many times the statistics are oriented primarily to show only unexpected situations and "no news" is assumed to mean "no problems". This type of "information environment" allows the decision-makers to concentrate on more acute problems. In this type of situation the information service is functioning literally as an alarm system. These examples indicate that the information producer cannot alone know what are the possible uses and how critical his production can be.

12. Communication and presentation

These two concepts can be discussed at very different levels. Communication in relation to systems - including health information systems - is the element which makes of autonomous separate units a "whole", a "system". "No communication - no system" presents the basic philosophy in its most simple form.

This aspect is really vital from the top management point of view. There is no need to discuss all possible methods of communication which can be used, e.g. at national level, in order to have the traditionally existing uncoordinated information sub-systems function in a way that can be called a "service".

The problems of communication must also be discussed at a more practical level. In the previous chapters some of the problems of close cooperation have been mentioned. Timing, exactness, etc., can only be agreed upon when both the users and the producers understand the realities. One of the problems which much be solved is the presentation of data in such a form that communication with the users is possible. Too much scientific jargon prevents the normal reader from seeing the essential facts. A list of all possible sources of errors helps the manager to be more critical; most likely, he will for safety's sake not use the information at all. Very often the human ability to read and understand a message presented only in numbers is grossly over-estimated. Some generalization and indication about where the essential findings are located, is definitely a prerequisite for any mutual understanding.

Well organized information service means much more than statistical tables; it means interpretation; it means that analyses and other knowledge, that leads to logical explanations, is readily available.

13. Lessons to be learned

Information services are such an elementary part of the health system that they can serve as an indicator concerning the situation as a whole. In principle the dilemma is at higher managerial levels: there is reason to suspect that the present generation of managers is not educated in systems analyses and will not see its benefits. Most likely the whole concept of modern management, the role of information, and the essence of information services, is not properly digested by the leading authorities. Obviously the same can be said of technical experts who do not have any common "language" to be used when discussing these kinds of problems.

The previous discussion is primarily focussing attention on the fact that information services for national health can be planned, must be planned and the responsible management has the leading role in this. Without their help the two most crucial problems - to have information which is needed and to organize the system in a purposeful way - cannot be achieved. Social organizations, including the health services, have a marked tendency to become more and more complex and only high level management can estimate the developments in the future. In principle, the information services are never fully finished - they require continuous rethinking, replanning, and especially retraining.

It is outside of this presentation to discuss which methods of training should be given, in which order, how much, to whom etc. A wait-and-see strategy, which is only based upon a future generation of more properly educated managers and statisticians is not a feasible one. In any case, the process is much too slow in the present situation. The training must be initiated now but the full results will probably be seen much later.

On the other hand a logical organizational structure or system makes management easier and more effective. To change ideas openly and to try in practice which seem to be reasonable is the recommended way to proceed.

14. Epilogue

Health is a goal which obviously never will be achieved completely. It is more a direction in which we should continuously be moving. One of the few aids which can compass us in the right direction is information services. They can intentionally be planned to serve such purposes. If compared with the cost of many other investments for health, information services for managerial purpose belong to the cheap ones. Some of the advanced medical technologies, valuable as such, help only a few patients with problems. The information services are helpful for everyone, including the healthy segment of the population. The introduction of a purposeful health information service does not require any radical changes in the habits of the people as a whole. Pragmatic applications of systems sciences serve well as a theoretical frame of reference.

The only thing which is required is some amount of interest, determination and sound reasoning by those who are responsible for the services, and the development of health policy.

References

1.   Häro, A.S. (ed.), Planning Information Services for Health, A Decision Simulation Approach, Health Service Res. by NBH Finland, Vol 22.,Helsinki 1981

RESEARCH AND DEVELOPMENT AND THE HEALTH SYSTEMS APPROACH

Detlef Affeld

Federal Department of Labor and Social Affairs

Bonn, Germany

Summary

Although health services research has seen a marked increase during the last decade, it seems to stem from different motivations and to follow different lines in the western industrialized countries. In general, the relevance of non-medical research in the health field for the development of health policy is very limited at the moment. Systems approaches play an even more marginal role in health services research and in health policy. Reasons can be found in general deficiencies about the social, economic and organizational aspects of health care, in the ambiguity of the system concept and the possible mix-up of intended and actual system coherence in the health field, and finally in the inadequacy of system approaches to cope with concrete peculiarities and given problems in historically specific situations of health care in different countries.
Health services research and health policy, in general, are largely concentrating on acute, short-term problems in a national or even regional or institutional setting. System approaches could be of value and add new dimensions to this, if they go beyond the traditional boundaries of national provincialism. Promising fields for the adoption of system approaches, therefore, could be

1. the possible general system dynamics of health care beyond the structural peculiarities of national delivery, financing, planning and policy patterns, especially the identification of strategically important options within health systems dynamics, and

2. the possible convergencies and divergencies in major structural variables of the health sector.

   Research and development in these fields has been extremely
poor till now, both nationally and internationally. System
approaches would gain in credibility if they started here.

1. Health services research in Western countries - common trends
   and different backgrounds

   Non-medical research activities in health care have been
rapidly increasing during the last decade. This can be shown
by the number of national and even international meetings and
workshops, by new research institutes and programmes as well as
by new professional journals. Even the health policy statements
of major groups involved in health, of political parties and of
government readily attribute high value to research, - in a
country like Germany and in other countries as well. This increase
is especially marked in the fields of social medicine and medical
sociology, in epidemiology, in health economics and in health
related organizational and political sciences.

   To give some general evaluation of ongoing research in these
fields is impossible here. But from German experience and from
the viewpoint of research policy administration, it seems worth-
while to mention some differences in this development between -
roughly speaking - the Anglo-American countries on one side and most
of the Western European countries on the other:

• In the United States, in Canada and Great Britain, non-medical
  health research has a scientific tradition of its own already. Academic interest in these questions and practical
  research have developped out of the scientific community for
  more than nearly 30 years. Actual developments like health
  care cost "explosion" and cost containment have only added
  specific accents and political relevance to this ongoing
  process. Health services research, in these countries, seems
  to be well established, and regularly funded at the universities.

• In Germany, France and the Netherlands, for example, health
  services research is most of all stimulated by the research
  demands and information needs of the administrative and political field. The scientific community itself does not show
  very much initiative of their own to enter into these
  questions. Research activities are largely dependent on
  special funding;university departments or institutes, if
  there are at all, have been founded very recently.

This very rough diagnosis might help to explain some
differences in the amount, the scientific quality and the choice
of problems tackled by researchers in the different countries.
Most of the methodologically interesting and more fundamentally
oriented health services research can surely be found especially
in Northern America. This seems not to be the same case with
regard to concrete problem, policy and action directed research.
American researchers and research administrators do not deny the
impression that the very existence of enormous data sources
(although to variable quality and with gaps) and the vast literature on social and economic aspects of health care does not
necessarily imply their practical and political utilization and
does, therefore, not prove the scientific sophistication of
American health policy.

In fact, it could be a scientifically interesting attempt
to comparatively evaluate the interdependencies of health
services research and health policy. Such a study, which could
be surely done under a general systems approach, would perhaps
not show a very good correspondence between any measure of system
coherence in national health services and health services
research. It might turn out, instead, that countries with a low
health systems coherence have a high intensity of health services
research and a broad variety of different scientific approaches
including a systems approach. A country like Great Britain, with
a higher degree of health systems coherence and a good tradition
in health services research, demonstrates a dominant tendency
to concentrate on management questions of single health delivery
institutions and a marked reluctance in the case of research
directed towards the health care pattern as a whole. Finally,
in a country like Germany with very low system coherence in the
health delivery services, research is concentrating on practical
and urgent information needs far more than on systematic and consistent professional approaches. The general systems approach does
not yield the needed short term assistance and, therefore, plays
a very marginal role.

2.   The actual role of systems approaches in health services
     research and in health policy

To attribute very marginal importance to systems approaches
in health care, at least at the moment, does not mean that it is
unknown or not discussed, - but the opposite. As a technical,
perhaps even more a vague political term, it has been common
to speak of the "health system" or the "statutory health insurance
system" in Germany for many years. To illustrate this:

- The term "system" is extremely widespread in health policy
  statements, discussions and journalism. It is unconsciously

used in almost all combinations. This use demonstrates
vagueness and imprecision rather than deliberate holistic
viewpoints.
- "Health systems research", as a technical term, has entered
  into the health policy statements of the German conservative
  political party as key term for research strategy in this
  field.
- Only a few years ago, two of the very few health services
  research institutes in Germany adopted this term in their
  official denomination.
- Some years ago, the large state-run research institutions in
  Germany (so called "Grossforschungseinrichtungen") set up a
  standing workinggroup on systems analysis. As part of this,
  they tried to do systems research in the health field too,
  without success. The attempt, therefore, was dropped very soon.

Health systems studies in Germany can be counted on the
fingers of one hand. And even this does not say anything about
the quality of the existing studies. The same picture might be
true for some other Western European countries. Leaving aside
the aspect of systems science as a discipline and turning to the
system coherence of given health care patterns as a practical
problem beyond one single scientific discipline, the situation
largely remains the same. By now, there is very few research concerning the "health system" as a whole or even concerning major
interconnections between large "subsystems". The hospital sector,
perhaps, with its ever increasing share of total health expenditure is one field, where this type of system questioning may
become a major guideline for research and policy very soon. The
interrelations between hospitals and ambulatory care as well as
general nursing care seem to be far more promising for cost containment than traditional one-sector hospital management
attempts.

3.   Reasons for marginal importance of health systems
      approaches

Primarily, reasons for the marginal importance of system
aspects in health services research and health policy might
be found in four fields, - with large differences between the
different countries:

- deficiencies in the general state of knowledge about social,
  economic and organizational aspects of health care;
- ambiguity of the system-concept when applied to the health
  sector;
- mix-up of intended and actual system coherence in the health
  field;

- inadequacy of the systems approach to cope practically with
  the concrete peculiarities of given health care patterns.

3.1. Fundamental deficits in empirical knowledge of non-medical aspects of health care

System approaches need the background of substantial, empirically based and tested knowledge with some explanatory value
from health economics, social medicine and medical sociology and
the health related organizational sciences. But these disciplines,
themselves, are far from any scientific maturity. Even in the
Angloamerican countries the hard core of empirically tested,
valid knowledge about the functioning, the prerequisites, the
effects and side-effects of given health care pattern or even
major parts of it are rather poor and, in most cases, open to
broad interpretation and speculation.

Besides this, only few of the Western countries offer the
type of well developped and sufficiently differentiated data
bases on health care functioning, financing, planning and decision
making that could be the starting point for interesting system
analytic approaches to the health sector.
In short : A systems approach to health care will surely be
seriously handicapped by the fact that it can not compensate for
the existing fundamental lack of the needed theoretical, empirical and data background for this field.

3.2. Ambiguity of the system concept in health care

To speak of system qualities or system coherence in health
care tends to leave open what aspects of the health sector
or what type of system coherence actually is meant or pursued.
To illustrate this : a low degree of system coherence in health
care delivery institutions may very well coincide with an
elaborated system coherence in health legislation. The German
example demonstrates this.
In the same way, high system coherence in delivery institutions
need not, by itself, be accompanied by the same system coherence
in health care financing. High system coherence in health related
political, organizational and planning processes can go together
with low system coherence in delivery institutions or financing
processes. Last but not least, high system coherence of health
institutions can be a major factor in low system coherence in
overall institutions, once the "system boundaries" have been
redefined even only slightly.
In short : The ambiguity of different possible and meaningful
"system"-aspects in health care makes it difficult to use this

concept practically with some value. It might be that the most
interesting and politically relevant questions are eliminated
just by the type of "system"-view favored.

3.3. Mix-up of programmatic and factual system contingency in health care

To attribute system qualities to the vaguely defined field
of health, could be premature, misleading or - in the worst
case - politically attractive semantics. To do so, might rather
express an idealistic view of some state of optimal and desired
organization of this sector or the wish for a normatively
loaded yardstick than an adequate approach to empirical analysis
of given services or a useful tool for stepwise reforms in
practical health policy.
Establishing the system concept as a key concept in health policy
and in health services research might - at least at the moment -
imply unseen or even deliberately unspoken a 'priori' decisions as to
the boundaries of the health sector and the priorities of health
policy. Inclusion or exclusion, for example, of collective
health prevention policy (environmental protection, housing
hygiene, improvement of working conditions, healthy nutrition
traffic safety, etc.) into the health sector, will result in
rather different views of the health "system" and a different
type of system-orientation in health policy.
System philosophy and system language in health services research
and in health policy may, at this moment, reflect the tendency to
strengthen the dominating curative and natural-science bound bias
of medical professionals. It does,perhaps, not point to more
prevention and more emphasis on psychosocial aspects beyond the
classical scheme of medical doctors' ambulatory and hospital
care.
In short : A premature and one-sided use of system concepts in
health care could help conservatism in health policy rather
than to analyze and overcome powerful interest structures.

3.4. Inadequacy of system approaches for the analysis of peculiar health delivery patterns

Foreign health experts have always been startled by the
extremely complicated organizational, financial and political
structure of the health sector in Germany. The so called
"highly structured statutoryhealth insurance" in Germany ("ge-
gliederte Krankenversicherung") with its 100 years of historical tradition has not ever undergone any fundamental reform or
revision. A huge set of additions, amendments and alterations
have made it the structurally dominant factor in German health

care. It has not so far been proved that system approaches
can adequately deal with the historic peculiarities of this
"system" without falling into pure trivialism. The very few
attempts to do so were not successful.
System researchers willing to enter this field with their
instruments, tend to give up once they have been confronted
with the absolute necessity to learn at minimum the leading principles of its highly complicated legal basis. Systems science,
in this respect, seems to be in a starting position like health
economics were in the beginning of the 1960's.

Health services research administrators, on the other hand,
do not expect practically useful and timely research results
from this approach. Unlike health economics, they therefore do
not foster such research lines by creating special possibilities
and funds for health systems research. It is not surprising
that this approach did not find any special mention in the governmental health research programme in Germany from 1978 or even in
the very recent European Community's research programme in
medicine and public health from July 1982.

4.   Strengthening system aspects in health care - where to
      begin with research and development?

The marginal importance of system approaches in health
services research and health policy is due to the predominant
tendency

- to handle non-medical health care problems in their very
  national or even regional setting as a type of unique
  phenomenon with almost no functional, structural and
  historical parallels in other countries;
- to focus on historically bound and very often "one issue"
  managerial particularities of health care;
- to expect practically useful remedies for limited problems
  in a short term perspective.

These tendencies are still very strong in most of the
Western countries. System approaches in the health field will
remain in a bad competitive position with other disciplines
involved and will keep their marginal role, if they do not add
a new dimension to the ongoing discussion.

It is striking, how much each of the national health care
"systems" is almost exclusively discussed, as if it were the
only one in the world. International comparison and systematic
evaluation does not go far beyond expenditure rates and input
statistics. The British and the German case, both being somewhat at different ends of a health care typology, clearly seem

to indicate this. One might even speak of some type of "contact anxiety".

The introduction of system aspects especially in health services research could and should fruitfully concentrate on overcoming the strong traditional boundaries of national provincialism in health policy. Health services research have, up to now, largely been bound nationally in their scope and perspective. They have not sufficiently, if at all, dealt with

- possible general system dynamics of health care beyond national peculiarities in delivery structures, financing mechanisms and policy procedures;
- possible convergencies or divergencies in major structural variables of the health sector in different countries.

4.1. Research in the general system dynamics of health care

The significant differences between, for example, the U.S., the British and the German health care "systems" can be interpreted as fundamental differences based on national peculiarities and starting from basically incomparable grounds. This seems to be the classic predominant way of thinking. A different and perhaps more promising interpretation could be to explain these differences on a background of very broad and similar system dynamics in health care. Different national "systems" could then be seen as passing through different stages of system development and stressing different aspects of a, nevertheless, rather similar system development process.

Throughout the literature of the last 25 or even more years, one will find many educated guesses and arguments that secular trends have a much higher influence on the development of the health sector than the diverse structural variables in different historic health "systems".
Taking this as starting point for system approaches in health services research, it would perhaps help better understanding about some of the involved questions like:

- What role do demographic factors play, especially the ever increasing proportion of elderly people, for the system dynamics of health care? Will this factor have a major structural effect on delivery systems, financing mechanisms and overall health care expenditures, which is more important than diverse structural starting conditions in given health care settings?

- How much do changes in the general panorama of health risks and diseases with an ever increasing proportion of chronical diseases affect the overall system dynamics of the health sector, apart from given structural differences? What do such

changes mean for the balance or imbalance of curative and
preventive aspects in health care? Will they foster psycho-
socially based medical paradigms besides the natural sciences
dominance? Will these changes significantly affect delivery
structures and manpower development whatever their actual
situation in different countries is at a given moment?

- Is the development of the medico-technical and the pharmaceutical complex the driving force in general systems
  dynamics of the health sector? Does its influence outreach
  the effects of classic organizational variables in the
  different countries?
  What consequences would this have for system steering, for the
  growth of health costs and the limits to their financing?

    Throughout the literature, too, there are strong arguments
that, from a certain stage of health systems development, the
classical health sector has a very small, if not completely
marginal importance for further improvements in the general
health status of a population.
Systems approaches could simply omit such questions by means
of restricting the health system definition to classical schemes
of health care delivery. It would, thus, avoid basic disturbances
and possible new insights at the same time. Keeping the system
approach within these limits of just new phraseology for well-
known questions is not very promising.
System approaches could, on the other hand, perhaps contribute
to a better analysis of just these central questions. And this
would suggest ways to overcome a stage of rather inefficient
system stagnation which most of the different national health care
patterns seem to be in. What consequences, for example, could
be drawn for the health sector and its future development into
a "system", if future improvements of the health status can be
primarily reached by a general political shift towards more collective prevention? Is this one of the really important strategic
alternatives and options in the system dynamics of health care in
different countries?
This leads to one of the most interesting questions within such a
system dynamics perspective of health care: What really are the
strategically important options in the development of better
health conditions? Have we already recognized and discussed them
carefully? Do we concentrate adequate research and political
attention on them? Or do we only follow worn out paths with just
new shoes under our feet?

4.2. Research in convergency and divergency of significant
       structural variables in different countries

    During the last decade, fundamental principles of the
health sector organization were the topic for controversial dis-

cussion. This happened in almost all Western countries regardless of their actual health care structures; it is largely due
to a worldwide health cost "explosion". Even countries with strict
health care budgets as the most direct and most effective means
to contain health expenditures were not exempted from these
deeply rooted disturbances.

As a side-effect and almost automatic consequence of global
cost containment policy, the athmosphere of insecurity thus
created has produced increasing willingness to redefine basic
principles in the general structure of delivery, financing and
decision making in health care. Some countries have already
basically changed their health policy. Italy is surely the most
striking example for such a radical change within short time.
Greece is undergoing similar developments. Less radical and not
always in the same direction, similar activities and attempts can
be found in the Netherlands, France, Canada or Sweden. And even
countries like the U.S., Great Britain or Germany, which are
traditionally resistent to fundamental discussion of alternative
health care organization patterns, are facing basic questions.
The German federal government, for example, during actual cost
containment legislation has recently announced that basic structural reforms of the health sector will be submitted to parliament in 1984.

Starting positions and directions of these structural
reforms in Western countries are ranging over a broad scale.
In some cases, countries are deliberately heading for organizational patterns, which other countries who traditionally have
experienced this envisaged type of organization, themselves
are eagerly willing to leave in the opposite direction. Furthermore, it is striking to see that on the national health policy
level very often such envisaged changes are characterized as an
apocalypse and a severe threat for the patients, regardless
of the performance and outcome in neighbouring countries with
just that devilish alternative health care pattern. Health
services research very often does not counteract this national
provincialism, but sometimes even follows it.
Till now, research and development have not seriously entered
into the questions of partial convergencies and divergencies
among important structural variables of different health care
patterns. Some major lines in this process should be mentioned
here:

- There are convergent as well as divergent tendencies in the
  role that state and /or self-governed, non-governmental bodies
  should play in health care. In Germany, this is one of the most
  important structural questions with far reaching implications
  for a wide range of problems in delivery, financing, planning
  and decision making.

- There are convergencies and divergencies as to the relative roles either of the private market or public administration in the health sector.
- The economically motivated need to discuss either limitations in the spectrum of benefits, which are offered by statutory health insurance or national health service, or to open up new forms of additional out-of-pocket financing ("co-sharing") seems to increase and grow universal and stimulates convergent processes in health care limiting and financing.
- There is a universal tendency in different health delivery structures to stop growth or even reduce expensive hospital care and organize compensatory facilities. This tendency favors convergent processes in the delivery structures in general.

The scientific analysis of these processes may be very suited to the adoption of system approaches and would offer a comparatively advantageous position for such a new approach. Do, for example, technically, scientifically and economically based structural variables in health care demonstrate apparent convergencies in system development? Are, on the other hand, divergencies largely the result of different interest and power structures dominating system development? The analysis of such convergency and divergency processes could perhaps help with the better identification of those fields that are governed by health policy and those by health politics.
Health systems research with this scope and purpose will almost automatically start with international comparison. International and supranational bodies like WHO, the European Community, NATO and others can play an important role to favor such an additional research strategy. In fact, they could be the only agencies with a good probability to overcome the inherent and already cited provincialism in health policy and research. They could be, - if they themselves were not following research strategies which result in the duplication of national research questions and in a high degree of pure description. The European Community with its recently announced second research programme in medicine and public health has had this chance; it has not been used very successfully so far and there is much doubt whether it will make the best of it in future.

III. NATIONAL EXPERIENCES

Albert van der Werff

Ministry of Health and Environmental
Protection, Leidschendam

National University of Limburg
Maastricht, The Netherlands

Introduction

The organization and development of health services
systems that has been conceptualized above are obviously
not easy tasks. They face many problems and constraints
in countries. These problems and constraints are both
physical and socio-political, and their size and nature
obviously differ among countries. Strengthening
the systems characteristics of health services can be
more successful, of course, if there is clear understanding
of the nature of the difficulties. The evaluative
studies on national health services and on different
sub-matters which follow can be used to identify many
of these problems and constraints.

THE FRENCH HEALTH SERVICES SYSTEM

J.F. Lacronique

Attorney of Mission, Ministry of Public Health

France

General Objectives

The French Health Care System is characterized by its am-
bition to conciliate two major ideological orientations, that
are classicaly considered as contradictory:
- the first one is egalitarianism, as it is expressed in the
  preamble of the French Constitution "The nation guarantees
  to all protection of health...";
- the second one is liberalism, since the vast majority of the
  physicians still work in private practice.

In order to achieve what may appear as an impossible compromise, the financing of health expenditures is assured through a
mandatory health insurance scheme, which now covers in principle
the whole of the French population.

However, it was stated in 1945 when this system was created,
that the Social Security in France will not merely work as an
insurance for health related risks, but also as a redistribution
of income mechanism, aimed at serving solidarity principles.
This last objective was supposed to be achieved by the fusion of
the various regimes applicable to the different professional
categories into a large and comprehensive insurance plan for
every citizen. But this unification was not even attempted, thus
leaving a very complex situation where traditional structures

survive, mixed together with more modern conceptions of health
care organization.

Political Structure of the Country

Since May 1981, the French Republic has been governed by a
coalition of socialists and communists, after a general election
that made the end of 23 years of "Gaulist" political majority.

The newly elected President, Francois Mitterand, is a socialist. His mandate at the Presidency is seven years. The new
Chamber of Representatives is also led by a majority of socialists,
and is elected for a five years mandate. Thus, the current Government of France is constitutionally assured to enjoy the power for
at least five years without fearing any foreseeable change.

During the first year of socialist Government, the accent
was essentially placed on a reorientation of economic policy,
giving priority to the fight against unemployment. To achieve this
objective, the Government placed its major thrust in stimulating
the general market by elevating the purchase power of the consumers. One illustration of this policy has been the distribution
of money to low-income families and the creation of new 150.000
civil servants jobs, thus expanding the public sector. This policy was in marked contrast with the former Government orientation
which was committed to reduce the role of the administration and
to impose "austerity" in economic life.

The expected outcome of the new policy was a general growth
in the consumption which would allow the private sector to invest
and create jobs. However, after one year of various incentives
aimed at convincing the private investors that the future was
promising, the Government realized, in May 1982, that the inflation
rate was becoming a major problem and decided to take drastic
measures to fight it.

In June 1982, a devaluation of the french money was decided,
accompanied by a three months freeze in wages and in prices. Despite an almost unanimous deception, these measures were approved
by the socialists and the communists. In summary, the general
climate of confidence to the socialist Government that prevails
since May 1981 has not been markedly affected by the poor successes that could be reported.

One reason for the lasting popularity lies in the fact that
the socialist Government can still blame the former Government
for its "heritage" of a very damaged economic situation. But
everyone can predict that such an argument is highly vulnerable,
and the real test of stability will take place in October 1982,
when the wage and the prices freeze will level off.

The Health System Structure

The French Health Care System is essentially characterized by its pluralism, that is to say that private and public services are available within the same insurance framework.

For a french citizen, this pluralism represents a very attractive feature, since it gives an almost total freedom to choose one's physician, or one's health care institution. This freedom is, of course, accompanied by some financial restrictions, and the patient is not reimbursed in the same conditions, according to what sector he chooses. But these inequalities are more illustrated by different types of coverage mechanisms rather than by different rates of coverage. For example, an out-patient visit in a public hospital is still at a very low price (23 F) because the patient pays only the co-payment part of it. In a private clinic, the patient has to pay the total fee, and then will ask for reimbursement.

Despite these differences, one can say that the private and the public sectors are in competition for certain types of care, like general and specialized surgery, obstetrics, rehabilitation, and long-term care for aged persons, disabled and mentally sick patients.

But for the usual medical activity, there is a marked specificity between the private and the public sector:
- The majority of out-patient care is assured by the private practitioners, working essentially in solo practice. 50% are general practitioners, and their distribution, however uneven on the territory, does not leave any ill-deserved areas. The doctors are paid on a fee for service basis.
- The majority of the hospitalized care is assured by public hospitals, which represent 70% of the total beds capacity. In these institutions the doctors are salaried. There exists also not-for-profit private hospitals (oncology centers and charity hospitals) and public municipality-owned medical centers (dispensaries), etc. Thus, pluralism in France shows not only a large variety of institutions, but also a very complex set of administrative status for each type of institution, mostly illustrated by different career development schemes for each category of professions.

The coordination of such a system is a very heavy endeavour, complicated by the multiplicity of ministerial tutelage. For example, the Ministry of Health is responsible for the setting of the management policy for the hospitals. But the Ministry of National Education is the leader in dealing with medical schools and training of professionals; the Ministry of "National Solidarity" is the only one to debate the fee schedule and the reimbursement rates with the different professionals, public as well as private.

Occupational health is supervised by the Ministry of Labour, foreign cooperation is under the responsibility of the Ministry of External Relations, etc.

In evident contrast with the dispersion of authority at the central level, there is a highly concentrated power at the local level: in each "Departement", the "Préfet" (called "Commissaire de la République" by a recent law) represents the Government, and must implement in real life all the regulatory decisions.

The "Préfet" is the head of a complete administration. Under his authority, a "Directeur Départemental des Affaires Sanitaires et Sociales" (DDASS) coordinates all the activity in the health and in the social sector, including Social Security. The administrative power of a DDASS is important, since he controls the budget of most of the health institutions, in accordance to principles set out each year by the Ministry of Health, the Ministry of the Budget and the Ministry of National Solidarity (Social Security).

The Production of Health Services

1. Coordination of Manpower Training

During the past thirty years, there has always been a controversy over the adequate number of physicians that was needed in the country. In 1955, the general opinion was that there was a surplus of physicians, since the newly installed doctors had difficulties in creating their practice. Then, ten years later, after the setting of a "convention", that made possible a governmental control of the fees demanded by the physicians, the demand for health services grew faster than the supply of physicians, thus leading to a situation of scarcity. Despite the relatively high flow of new doctors issued by the french medical schools in 1965 (5.000 per year on average) the Government opened several new medical schools and conducted a policy of rapid growth of medical manpower. By the year 1972, the capacity of the medical schools reached the level of 10.000 new doctors per year.

Then, in 1975, the fear of a surplus reappeared, when it became clear that the current flow of new doctors would lead to a total supply of more than 200.000 physicians by the year 1995 (1 doctor per 300 inhabitants).

The Government, thus, decided to decrease gradually the capacity of the medical schools. The initial objective was to lower down to 6.000 per year, that would lead to a steady situation of 170.000 doctors by year 2000. In 1980, it became apparent that the growth of the number of physicians was among the causes of rapid

growth in health expenditures, and thus, the Government decided
to push down the capacity of the medical schools to 5.000, to
show its commitment to a long term control of health care financing.

This policy was called one of "rationing" by the political
opposition, which fiercely criticized it when it was presented to
Parliament.

Thus, when the Socialists came to power, it could have been
expected that a new period of expansion would be opened up for
the medical schools. However, the new Government decided to avoid
any brutal move, and "freezed" the flow of students admitted in
the second year of medical school at the level it had reached in
1981, that was 6.500. A similar policy applied to dentists, pharmacists, and nurses, however less dramatic than for the physicians.

In conclusion, one can infer from this account of the recent
evolution, that a control of the demography of the medical professions is working in France. There is now a political consensus
that a severe limitation has to be respected.

But a merely quantative control is not sufficient, and a
complete reform of the medical training is currently undertaken,
which will place the french system in complete harmony with the
other E.E.C. countries. This reform is placed under the double
tutelage of both the Ministry of Education, and the Ministry of
Health. Its principal feature is to improve the practical training
of the student by providing a minimal hospital internship of two
years for G.P.'s, and of four years for specialists.

2. Coordination of Hospital Equipment

In 1970, a major law concerning the hospitals was passed in
France. According to the provisions of this legislation, the
Government would have control over the creation of the modifi-
cation of any hospital beds, and of any costly equipment. The
law provided that an inventory of all the equipment would take
place, and that "standards" and "ratios" would be set up in
order to achieve a better distribution of the facilities and of
the medical equipment all over the country. This procedure was
called the "carte sanitaire", and worked quite well in achieving
its role as a rationing device. For example, France is, in Europe,
the country where the number of scanners is the lowest, because
the law provided that the adequate ratio for that type of equipment was one per million inhabitants (this rate was recently
changed to a new ratio of one machine per 600.000 people). There
exist such ratios for kidney dialysis equipment, for telegamma
therapy, for openheart surgery machines, etc. However this regu-

latory mechanism faces strong criticism, for it prevents sometimes
the diffusion of new equipment that is often judged by the physicians as "necessary" for the patient. For example, a recent mono-
graphy was published, accusing the Government of being responsible
for the death of hundreds of patients per year, due to a lack of
3.000 artificial kidney machines.

3. Coordination of Research and Technology

Research and technology were placed, in the past, under the
authority of the Ministry of Industry. Then, ten years ago, a
special agency was created, under the direct authority of the Prime
Minister (Délégation Générale à la Recherche Scientifique et
Technique). The new Government of France includes a full Ministry
of State for research and technology, thus illustrating a strong
commitment to foster Research and Development. The objective is
to increase the research budget from 1,2% of the G.N.P. to 2,5%
by year 1983, under the assumption that R. and D. can be a solution
to the economic crisis, and should not be its victim.
At the same time, the Government has also expressed a concern for the evaluation of the consequences of modern technology, and has created in March 1982, a "Center for the Assessment
of Sciences and Advanced Technologies" (Centre d'Etudes sur les
Sciences et les Techniques (C.E.S.T.A.), which will work on
an inderdisciplinary basis.

The Economics of the Health Care System

The consciousness over the excessive growth of health expenditures rose in 1975, when Madame S. VEIL was Minister of Health.
One of the first steps in tackling the problem was to place Social
Security and Health under the same responsibility. This decision
was aimed at showing that "Health" could no longer be a pure "expenses sector" but should be considered together with its resources
funding mechanisms.

The integration of a single agency to coordinate the Health
and the Social Security services worked quite successfully for
five years. However, it did not lead to any leveling off of the
health expenditures, that continued to grow at a rate of 18% on
average. In august 1979, the Government decided to make a set
of drastic decisions to control the rise of health expenses, and
indexed any future increase in the health budget on the growth
of the G.N.P. This policy was not well received by the Health
professionals, who consider themselves as a priority sector,
that justifies a higher rate of growth than the one of the National
resources.

In May 1981, the new socialist Government stated clearly
that it would not follow the "austerity policy" of the former,

and decided that the entire social sector should be considered as "social investment" priority sector. At that time, it was possible to display a certain contempt of the economic consequences of this policy: in July 1981, the Minister of National Solidarity said publicly that "she would not be a Minister of accounting", just to show that she would not be afraid of spending more than the resources would allow. Few months later, she recognized that she should not have made this declaration.

a/ Financing Health Care

Most of the money used in the health system in France comes from reimbursement by the Social Security. The few exceptions are illustrated by some programs funded directly by grants from the Ministry of Health to take care of special categories of patients, i.e. infant and maternity care, ambulatory psychiatry, prevention and immunization programs, etc.

But the general rule in France is that the health expenditures have to be covered by the contributions that all the active workers give to the Social Security Fund. Thus, any increase in the expenditures must be matched by a corresponding increase in the contributions.

The rate for these contributions is fixed by the Ministry of National Solidarity. In 1981, an increase of one percent was set to balance the social security budget, thus placing the average contribution to the social security of a French worker at 19% of his total revenue (the rate of contribution varies with the salary, and is spread on both the employee (1/3) and the employer (2/3).

The money is collected through a special fund that exists in each "département" (U.R.S.S.A.F.), and is then redistributed to the different agencies that are specialized in the provision of a given service (retirement plans, maternity care, family compensation, sickness, unemployment, etc.) The sickness fund represents 35% of the total social budget.

National health insurance does not cover all health expenditures. There is still a co-payment (ticket modérateur) which can be supported either by out-of-pocket money or by a special semipublic and volontary mutual-benefit fund, or a commercial insurance. There are about 8.000 mutual-benefit societies, the majority of them being professionally oriented. About 60% of the French population subscribes to this form of voluntary insurance. In 1979, the Giscard d'Estaing Government tried to force these mutual-benefit societies to implement a small co-payment (Ticket moderateur d'Ordre Public (T.M.O.P.), in the hope that this measure would make the consumers more conscious. This decision was ill-received and stirred a violent opposition, until it was withdrawn in 1980.

Finally, where social security covers most of health expenditures, the coverage varies greatly according to the type of service that is provided. Table I shows the source of finance for different types of services in 1978, in percentage.

TABLE I

Sources of Financing in % in 1978

Source of financing Type of service	Social Security	Mutual Aid Societies	Consumers	Public Administration
Hospitals	88	1,3	6,1	4,4
Physicians	63	6,5	30	1,5
Dentists	36	4,7	60	0,2
Pharmacy	56	6,3	36	1,6
Total	71	3,8	22	2,8

b/ The Health Care Services Consumption

In 1981, the French population spent 7,8% of its G.N.P. on health care. Six years before, health expenditures represented only 6,7% of G.N.P. This rise signifies an average increase of 16,8%; about half of this rate can be accounted for to a rise in consumption, the other half is due to a rise in prices. However, the different types of health services did not follow the same rate of growth. Table II shows that hospitalization has been by far the most important factor of growth in health expenditures.

Table III shows the allocation of expenditures among different types of health services in 1980.

However these figures do not reflect one characteristic of the pattern of consumption, which is a striking geographical variation. The disparities in medical consumption can range from 1 to 4 in hospital-days, in medical-technical procedures, in

TABLE II

Evolution of the Consumption of Health Services
(rates of growth between 1970 and 1980)

	1970	1980
Ratio Total Health Consumption/ Gross National Product	5,7	7,5
Growth of the Total Health Consumption . Total .... adjusted for price increase..	17,2 8,8	18 3,2
Growth of G.N.P. . Total .... adjusted for price increase..	13 % 3 %	
Growth by types of health services (1970-1979) . Health insurance budget...... . Pharmacy..................... . Physician's fees............. . Hospitals....................	18 % 12 % 16,3 % 22 %	

TABLE III

Health Expenditures in 1980

Type of service	Percentage
Hospitals.........................	50,2 %
Pharmacy..........................	18 %
Physician's fees..................	14 %
Dentist's care....................	9,5 %
Biology lab. tests................	2,3 %
Glasses, aids, etc................	1,9 %
Total : 3.836 F per person	

home visits. The most important factor that explains these
variations is the provision of health personnel and facilities.

The heterogeneity is also the "price to pay" to the total
freedom of location of practice that the French practitioners
enjoy. So far, there has been no attempt to set any regulatory
mechanism to solve this question of distribution of services,
with the exception of the "carte sanitaire" which concerns only
hospital beds and costly medical equipment. The only initiative
taken recently by the Government has been a purely informative
system that makes statistics available to a physician looking
for a place to settle and establish his practice.

However, the new communist Minister of Health, Jack RALITE,
has expressed his intention to make efforts to correct these
disparities. He likes to refer to "unequal measures to right
inequalities", and has already illustrated this position in
allocating the newly created positions in school preventive
medicine, to underserved areas, instead of spreading them over
total territory.

c/ Recent Political Orientations of the French Health Care System

Under the pressure of the change in medical knowledge, the
different medical professions were forced to waver between making
progress more accessible (especially in the public sector) and
protecting liberal medicine. A subtle balance between public and
private initiative so far has been preserved, in almost all the
types of services. In May 1982, the new Government re-affirmed
its commitment to respect pluralism, in publishing a general declaration on health, the "Charte de la Santé". In this document,
the Ministry of Health commits itself to encourage the diversity
in the provision of health care.

In the ambulatory care sector, it guarantees independance and
freedom to the private practitioners. It announces also that new
experiences, like "community health centers" will be encouraged.
In the hospitalization sector, a priority will be given to "general
hospitals" (non-teaching hospitals). A reform of the hospital professions is scheduled, on the basis of a unique statute for all
the salaried full-time staff physicians.

Finally, the Charte de la Santé insists on the necessity
of controlling the health expenditures saying that "the health
system must make accounts to the nation. Any wasted money is a
precious resource that is lost for the community, and also
hinders an effort to healing someone, somewhere...".

In order to achieve the objective of control, the French
Government will develop instruments for evaluation, in the public

as well as in the private sector, and will implement a budgetary policy for hospitals.

In terms of concrete changes, there has been little transformation into the functioning of the system. The major decision that has been made until now is the withdrawal of the authorization that full-time doctors had to see private patients in the public hospitals.

But various projects are currently under way and will undoubtedly raise some controversies next year. The most important reforms that are currently in discussion deal, (and will then be presented under the form of a law), with the following issues:
- Emergency care (Loi sur les urgences médicales)
- Post-doctoral training (Loi sur la réforme de l'Internat)
- Mental health (Loi sur le secteur psychiatrique)
- Hospital administration (Nouvelle Loi hospitalière)
- Research administration (Statut de l'INSERM)

A new style of management has also emerged, which is characterized by a large participation of consultants that come from the trade unions, and the associations.

The Future of the French Health Care System

It would not be fair to forget that a large majority of the medical community was far from satisfied by the former Government. The french health care system relies to much on an impossible compromise between state intervention and liberalism, and is thus very sensitive to any political instability.

It is indisputable that a large majority of the general public likes the current type of organization of the health system. But the professionals are highly sensitive to the constant decline in their purchase power (- 6% per year in the past two years for the private practitioners). The recent freeze in wages and prices occured only two days after a promise made by the Ministry of National Solidarity to raise the doctor's fees, and thus stirred a new dissatisfaction.

In the salaried sector, there is also a lasting grungy mood, due to the lack of personnel in the hospitals. An effort to hire about 15.000 employees in the hospitals has been partly offset by the new measures reducing working time.

Thus, troubled waters are still ahead for the French Health System, in a general framework that is supposed to respect its currect structure.

THE DUTCH HEALTH SERVICES SYSTEM: PROBLEMS AND MEASURES OF INTERDEPENDENCIES AND CONTROL

A.C.J. de Leeuw
University of Groningen

I.M. Mur-Veeman
University of Limburg

The Netherlands

0. Introduction

To present a complete picture of the health services system of even a small country as the Netherlands is impossible. Perhaps the first and most impressive characteristic is the overwhelming complexity of the system. This system was never designed and introduced as a whole but is rather the result of a variety of mostly uncoordinated initiatives and measures. In the Netherlands many of these initiatives came from practising professionals and private institutions mostly of a religious background. This complexity asks for a global description of the main structural and cultural characteristics of the system, which is what we intend to do in this paper.
Our attention will be directed to the interdependencies in the system, trying to understand the system as a whole.
The problems associated with this system might be put into three categories:
- The system is considered too costly. It becomes unaffordable. Health services consume too big a part of the national income.
- The system is considered to be incoherent. This is not only supposed to lead to unnecessary spending, but also results in maladapted and uncoordinated treatment of the patient or, to put it in another way, the servicing of the client.
- The system is considered to be uncontrollable.

It is believed that these problems can and must be dealt with: so there are strong tendencies for immediate reduction

of costs and for structural changes of the health services.
In this paper we will concentrate on the structural aspects
because they are central to the above mentioned problems.
But the financial aspects will also come up for discussion.

In the first part of the paper some concepts are introduced which we need for description and analysis. They
mainly are taken from systems and control theory.
We then describe in part 2 the structural aspects of the
system and its lack of coherence.
Part 3 is devoted to the measures being proposed and taken by
the national government to tackle the problems. There are
however strong resistancies and mechanisms which might result
in a worsening of the situation. Government policy might be
counterproductive. We point at three relevant factors: the
existing power structure of the system and the related personal
and institutional interests, the lack of coherency of the
governmental policy measures which are in conflict with each
other and lastly misconceptions about the appropriateness of
the policy measures.

Part 4 presents some measures which may provide positive
results: real decentralisation, restructuring of the financing
system and change of the educational system.

Finally some research themes are suggested.

1. Some concepts from systems theory

As our concept of system we take a system as a set of
elements with interrelationships, connected to the environment through input- and output-relations. The set of relations
together form the structure of the system.

In a system we can discern three types of partsystems:
subsystems, aspectsystems and phasesystems.
- Subsystems can be taken from any system by identifying
  some subset of elements of the original system and all relations which exist between those elements. For example
  the laboratory is a subsystem of a hospital.
  Other examples are the surgery unit and the board (i.e.
  the formal governing body of the institution).
- Aspectsystems are taken from the original system by taking not
  all elements but only certain relations (aspects). One can
  for example study the information flows, the power relations
  and the financial relations. These are aspectsystems.
- The concept of a phasesystem relates to the temporal dimension
  of a system. One might take out of the original system some
  specific period of time to get a phasesystem. This could be
  done to study shorttime behavior or long-term developments.

The concepts of structure now can be enlarged. We not
only have relations between elements but also relations between
partsystems.

Complexity is mostly defined as the number of elements and
relations. We however prefer to define complexity as related to
the problem of control. This can easily be understood if one
realises that we talk about complexity because the system cannot
easily be understood and controlled. Complexity thus refers to
the difficulty of control.
Complexity then might be related to decomposability (a decomposable system is less complex than a system which cannot be
decomposed), controllability, predictability and the information
that must be processed.

Let us go further into the concept of decomposition. H.A.
Simon (1962), in his famous article entitled "The architecture
of complexity" demonstrated that complex but viable systems have
a hierarchical structure. That is, they consist of weakly related subsystems which consist of weakly related subsystems
which consist of .... until we reach the smallest element.
The fact that the relationships between the subsystems are
weaker than within the subsystems makes the system nearly decomposable. One can study and control one subsystem in depth and in
detail without paying much attention to the others. The relations
within any subsystem produce relatively fast short-term behavior.
The (weaker) relations between the subsystems produce the slower
long term global behavior.
In studying this one can neglect details within subsystems.
It is easy to understand that the total control problem of such
systems can be split up into manageable parts. The viability
of such systems appears also if one realises that they have
an ultrastability property (Ashby, 1960): if one of the subsystems is disturbed the weak couplings with other subsystems
prevent the disturbances to affect other subsystems. Hierarchical
systems are much more stable. If the couplings between subsystems
are relatively strong, the control of one subsystem may introduce
disturbances in another one, so the control has to be coordinated.
Weak relations between and strong relations within subsystems
mean that there are clear and strong boundaries around the
subsystems.
Looking for the complexity of a system one can thus study the
boundaries between the subsystems. If clear and strong boundaries
cannot be found the system is not very stable and very difficult
to control.

Elsewhere the concept of a hierarchical system in the Simon
sense is broadened (De Leeuw, 1980). Instead of a system consisting of relatively autonomous subsystems and so on, we define
a hierarchical system as a system which consists of relatively

autonomous partsystems which consist of relatively autonomous partsystems and so forth. This enlargement means that
one can also look for boundaries between aspectsystems and
phasesystems. This is not only of an academic interest. The
fact that for instance the total management function can be
split up in functional fields of management is precisely based
on weak interactions between these functions or aspects. The
more relations there are between them the more they must be
coordinated and the less it is meaningful to split them.

Decomposability, hierarchical structure, stability and
controllability are thus strongly related concepts.
We may note that hierarchical systems are decentralised systems.
The higher level controls are of a broad nature. They leave
room for decisions in the lower level partsystems. Decision
power thus is distributed through the levels of the hierarchy.
If however some higher level controller (e.g. the central
government) tries to control the details within lower level
subsystems, the hierarchical structure is disturbed and at the
same time favorable properties are thrown away. Complexity
increases and stability decreases.

With this rough and incomplete sketch of some aspects of
complexity, hierarchy and control, we have argued that hierarchical
structures have very favorable properties and that they form a
key-factor for the reduction of complexity. We might add that the
argument can be made much stronger by introducing more analysis
of f.i. Ashby (1960), Koestler (1979), Simon (1962) and Conant
(1974 en 1976).

There is a last point we wish to make. If we can discern in
a system relatively autonomous partsystems (clear and strong
boundaries between them) we need little coordination. For only
the weak interrelations call for control. We have partsystems
S1 and S2 and weak interactions I. Control can now be dealt with
by a controller C1 and C2.

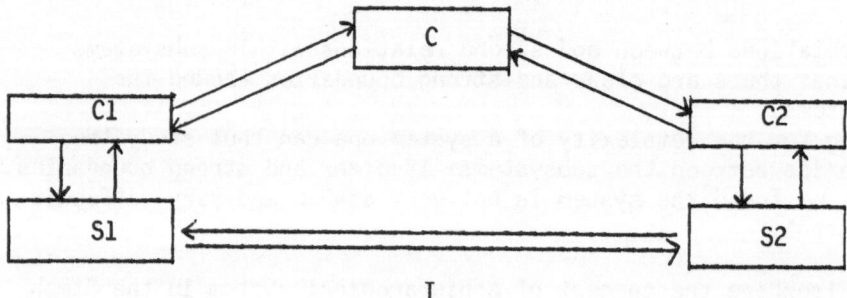

Figure 1.

There is a controller C to coordinate C1 and C2. C's task is
small and easy because I is weak.
When I is stronger, than C's task is heavier. Controller C
takes care of the coherence.
We thus must notice that:
a. It is important to look for the most clear boundaries.
   If we don't, the control task becomes unnecessary complex.
b. C1 and C2 must correspond with S1 and S2.
   If we don't stick to this rule control becomes unnecessary
   complex.
c. Controller C must not control the details of S1 and S2. Otherwise the complexity grows needlessly.
d. There is a balance of interactions between the partsystems on
   the one hand and the needed control to provide for coherence
   on the other hand. If I = 0, then coherence is assured without
   C.
By this analysis is demonstrated that the complexity we meet in a
system has to do with the structure of the controlled part
(whether there are weak boundaries or not) but also with the way
we organise the control.

   In the next part of the paper we will argue that these factors, among others, tend to create the problems we mentioned
earlier.

2. The Dutch Health Services System: Interdependency and Cohe-rence

2.1. Interdependences

   If the Dutch health services are described from the systems
point of view, the following picture arises: a very complicated
collection of partsystems, all overlapping and engaging with each
other, within a system which itself is a part of a very complicated society and which is bound to society in every possible
way.
As in other western countries the Dutch health services system is
fed by society through legislations, input of knowledge and
technology, input of labour and economic support (Field, 1973).
In exchange for this the health services system performs services
(or at least the system has to perform services), directed to
health maintenance and health-recovery of all members of our
society.

   The complexity of the health services system is caused i.a.
by the fact that the partsystems mentioned before can be defined
in many different ways, according to different criteria. However,
clearcut limits between these partsystems are hardly to be found.
Following the preceding analysis the system can be seen as extremely complex (for it is not decomposable).

The following distinctions can be made according to different
criteria:
- The echelons, including institutional and ambulatory health
  care.
- The sectors, such as general health care, mental health care and
  so on.
- Other classifications are also possible, for instance according
  to target-groups like aged people, children and youths, drug-
  addicts and so on.
- The professionals, like general practitioners and specialists,
  para-medical manpower, nurses, etc.
  Most of these professionals are united in professional organisations like the Royal Dutch Society of Medicine (KNMG), the
  National Society of Specialists (LHV), the Dutch Society of
  Privately Practicing Physiotherapists (NVVF) and many others.
  These professional organisations stand for the interests of
  their individual members in the first place. They sometimes have
  an advisory function to the government, while they also enable
  their members to exchange knowledge and build up relationships
  within their own profession.
- The public authorities, semi-public authorities and private
  institutions. It is possible to discern three levels of public
  authorities, namely the State Government, the Provincial Government and the Municipalities. Above all so far the State Government has paid attention to controlling the health services system through designing and enforcing legislation, regulations,
  measures and permissions. Now, through the planned decentralization and regionalization, the State Government seems willing
  to transfer a part of its controlling activities to the provinces and municipalities.
  We shall return to this subject in the next parts of this paper.
  The special advisory committees of the government like the Na-
  -tional Health Council, the Hospital Facilities Council and the
  Provincial Health Councils are considered to be semi-public
  authorities.
  Health care in the Netherlands is an activity of private institutions, as a rule foundations.
- The levels of aggregation.
  This concept refers to the distinction which can be made between
  persons, groups, organisations, functions of organisations and
  so on.

In the actual practice of health care delivery it is impossible to distinguish the same clearcut limits between various
partsystems as is done in the preceding description.
On the contrary, these partsystems are narrowly intertwined,
bounded to each other in every possible way.
Thus, within the echelons one can distinguish sectors, institutions or groups and the other way round. So elements of different

partsystems, for instance certain professional groups, authorities, institutions, etc., can form new partsystems, like
advisory committees or health councils.
Considering the functioning of such a committee or council,
we can notice that the distinguished levels of aggregation are
not to be clearly limited: for instance, a council-member
participating a titre personnel (at personal level), may function
as a representative of a certain institution (a hospital, a municipality).

If one tries to control the whole system, without realising
the interactions between the partsystems (the low decomposability), then an incoherent policy will arise.
If one realises the existence of these interactions and if one,
nevertheless, wants to control the whole system, then the control-
problem will increase to such an extent that controlling capacities will be failing.
If the control-problem is defined in such a way that it exceeds
the controlling capacities, incontrollability is the result.
Too little controlling capacities does not mean less goal achievement, but rather no controllability at all.
These interdependencies are shown in Figure 2.

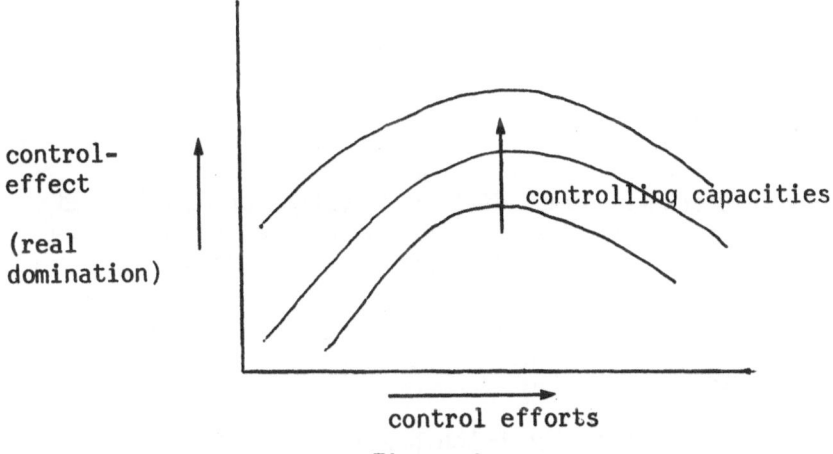

Figure 2

Of course we do not know the exact relations between these
factors, but there are some indications of its rough forms. It
is obvious that an increasing control effort will be accompanied
by a decreasing control-effect.
As far as the system nevertheless possesses some decomposability-
qualities, attention must be paid to achieve the right correspondence between control and decomposability as is suggested in
part 1. We doubt whether the Dutch health services meets this
requirement.

2.2. Lack of Coherence

It is remarkable that, despite the suggested narrow interdependency between diverse partsystems within the Dutch health
services system, these partsystems appear to function in a very
autonomous way and are even striving for autonomy.
Following figure 1 there is no balance between I (the interactions) and C (the controller who must guarantee the coherence
by coordination).
Leenen (1979) speaks of exaggerated aspirations for autonomy
everywhere in the health services system. He points to the fact
that health care in the Netherlands has not grown to a coherent
whole, partly because of its origins in the system of free practising individual professionals. The actual coherence and the
mutual influences are rather more the results of mutual effecting
and relatively autonomous forces than of a conscious decision
making policy. A total framework of a decision-making-system
is lacking. For this reason dominating and controlling the whole
system becomes a very difficult task.
Coherence can also be assured by a pattern of relationships,
especially arranged to secure coordination. We call this intrinsic control. In that case there is no need of a separate controlling body (extrinsic control).

The Financial System

We wish to point out that this lack of coherence can also be
found in the financial system of the health services. There are
different insurance-systems, different systems of payment and
different financial sources. The following brief survey of our
financial system is derived from Schrijvers (1980) and from Rutten
and Van der Werff (1982).

Insurance-systems

About 70 percent of the Dutch population is insured compulsory with the Sickness Funds and 30 percent (individuals with
an income above a certain level - DF1. 43.000,-- in 1982 -; self-
employed and civil servants) have to acquire private insurance
against health care expenditure.
The Sickness Funds are operating the social insurance system of
the Sickness Funds Council. Control is exercised by (representatives of) organizations of employers, employees and doctors.
The premium of the compulsory public insurance is a uniformly
fixed percentage of personal income (in 1982: 9.2 percent with
a maximum of approximately DF1. 3.400,--) and is met jointly by
the insured employee and by the employer. Privately insured
persons pay a nominal premium. Insurance for the "heavier risks"

i.e. for exceptional expenses, such as those incurred by patients
in institutions which care for long-stay patients and the physically or mentally handicapped, is governed by the General Special
Sickness Expenses (Act). The General Special Sickness Expenses
Fund was estabished in 1968 to cover these long-term and expensive costs.
In fact this scheme is a national insurance scheme, in principle
applying to all residents of the Netherlands. The premium is a
percentage of a person's wage or income (in 1982: 3.3 percent)
and is paid by the employer.

Systems of Payment

The independent practising professionals as well as the
ambulatory services and hospitals acquire their income by means
of a system of tariffs, a system of budget-financing or a mixed
system of both tariffs and budget-financing. It is noteworthy
that there are three tariff-units: a fixed price per bedday, per
consultation and per performance of a service. Hospitals are
dependent on the reimbursement of a fixed price per bedday.
Hospital boards and managers have to prove that the tariffs they
charge meet the actual costs before they get approval by the
Hospital Tariffs Agency.
This price regulation is based on the Hospital Tariffs Act, 1965.
Thus hospitals are dependent on the prices per bedday. But they
are also dependent on the specialist, because he determines the
extent of activities per patient.
The specialist gets a fixed amount of money for each publicly
insured patient referred to him by a general practitioner (who
issues a so-called referral card) for which amount one month
of outpatient treatment (consultations and minor operations) is
given. If continuation of outpatient treatment is necessary an
additional amount of money per month is given to the specialist.
For major performances in an inpatient or outpatient setting the
specialist receives a fee for service. In the private sector
specialists demand a fee for each service to their private
patients.
The institutions and professionals in ambulatory health care
are involved with many different financial systems. The general
practitioner is paid through a capitation system: he gets a
fixed amount of money per year for each publicly insured person
in his practice, regardless of the quantity of care provided
during that year. In the private sector he demands a fee for
each service to his private patients.
Until 1980 the institutions for ambulatory care, such as the so-
called Cross-organisations and some institutions for ambulatory
mental health services were dependent on subsidies from the
public authorities. A recent development is to finance their
services from the General Special Sickness Expenses Fund.

Financial Sources

It is possible to distinguish five different financial
sources, namely:
. The State Government.
  Transfers to the social insurance funds, contributions to
  academic hospitals and subsidies to a few institutions of
  ambulatory care.
. The Provincial Government.
  Subsidies - for instance for a limited number of integrated
  primary health care centres - and the cost of policy-making at
  the provincial level.
. The Municipalities.
  Some municipalities contribute to the costs of municipal hos-
  pitals; all municipalities pay for the municipal health ser-
  vices and preventive care.
. Families and Individuals.
  All families and individuals have to pay their premiums to the
  public or private insurance funds as mentioned before.
. Organizations (industrial and civil services).
  These organizations spend every year a certain amount of money
  on industrial health care services and on the protection of
  food and drinking-water.
Schrijvers noticed an increase in the contributions from the
State Government and from families and individuals during the
period of 1953-1974 and a decrease in the contributions from the
other financial sources.
This phenomenon is shown in table 1.

Table 1. Financial Sources in the Dutch Health Services System

	1953	1974
Sum in Dfl.	770.0	14.771.3
State Government	9.5%	12.9%
Provincial Government	1.5%	0.1%
Municipalities	10.6%	2.0%
Families/Individuals	76.1%	83.4%
Organizations	2.3%	1.6%

Looking at our financial system we can conclude that its complexity is very high:
- the interdependency is very high. It can be described as an
  untransparant muddle of financial flows (not-well decomposable;
- the predictability is very low. For there is an overwhelming
  amount of variables, which influence cost and expenses;
- the controllability is low. Prices are fixed by automatic
  couplings and norms.

Besides, there is a scarcity of control measures. The fact that the control system does not correspond with the financial system can be considered as an attendant problem. For instance, it is not unusual to speak about costs as the sum of all kinds of financial flows, but the summing up of all costs reduces the controllability of the financial system.
Moreover it is noteworthy that the choosen control-instruments (standardization, admissions) do not stimulate cost-saving behaviour.
The division between ministries leads to a lack of coherence. Thus we have to conclude that the system does not meet the requirements we suggested in the preceding part.

Unbalanced growth

The lack of coherence in the Dutch health services system is expressed not only by a dispersed financial system, but also by an unbalanced growth of the various partsystems.
Until the beginning of the seventies institutional health care was able to expand almost without limit; on the other hand, preventive care and ambulatory health services developed insufficiently. Further, a short-term narrowing of this gap is not to be expected, despite the fact that authorities as well as private institutions aim at strengthening ambulatory and preventive health care at this moment.
For this reason Leenen (1979) has stated that at this moment we are thinking in terms of ambulatory and preventive care, while policy making and the financial system do not support these conceptions.
This conclusion can also be derived from the figures in the so-called "Financial Survey", which is published each year with the annual budget of the Ministry of Health and Environmental Protection.
Some of these figures are shown in table 2, which is taken from the Financial Survey in 1981.

Routine-control

Finally we wish to point to the unintended effects of routine-control (i.e. controlling the existing planning and financial structure) on the health services' structure.
An example:
The "Memorandum on the Structure of Health Services" published in 1974 by the Ministry of Health and Environmental Protection, argued for regionalisation; the institutional and ambulatory health services had to be coordinated by region (coherence!).

A.C.J. de LEEUW AND I. M. MUR-VEEMAN

Table 2. Estimate of cost of health services in the Netherlands, 1981 – 1985   (in million Hfl.)

	1981	1982	1983	1984	1985	Changes in percentage with regard to the previous year				
						1981	1982	1983	1984	1985
Inpatient care	17.396	19.001	21.022	23.257	25.738	6.8	9.2	10.6	10.6	10.7
Specialist care	1.852	2.010	2.228	2.470	2.470	3.9	8.6	10.8	10.9	10.9
Pharmaceuticals	2.802	3.010	3.263	3.540	3.844	6.3	7.4	8.4	8.5	8.6
Outpatient care	4.745	5.232	5.837	6.475	7.176	5.8	10.3	11.6	10.9	10.8
Collective preventive care	793	875	981	1.100	1.234	7.7	10.3	12.1	12.0	12.2
Policy and Administration, ambulance service and other health services	1.985	2.190	2.443	2.718	3.020	7.8	10.3	11.6	11.2	11.1
Sub sum total	29.573	32.318	35.775	39.559	43.753	6.5	9.3	10.7	10.6	10.6
Economies 1982		210	234	260	305					
Reducements of super-norm-wages of privately practising professionals	110	285								
TOTAL	29.463	31.823	35.256	39.014	43.163					

A structural change like this can lead to enlarging decomposability (more independent regions) and increasing controllability and stability of the whole. To remedy the increase in
health expenditures,however, the policies are directed to introducing more governmental control (routine-control), in other
words decreasing decomposability.
In this way structural aims are opposed by routine-control.
Another example: the phenomenon of bureaucratic control mechanisms which function as learning mechanisms. This means that
management, functioning in the frame of bureaucratic structures
and bureaucratic control measures, tends to behave more and more
in a bureaucratic routine and rigid manner. So the system continuously is producing bureaucratic managers. This development
runs counter to a policy of increasing the quality and flexibility of management.

3. Mechanisms stimulating cohesion, dangers and counterpressures

3.1. Measures to stimulate cohesion

We cannot deny at this moment that efforts are being made
to unite autonomous subparts in the Dutch health services system
into a total interconnected system, with aspirations to better
controllability and reducing costs, which in the Netherlands, as
in other Western countries, have increased very rapidly (from a
share of 4.5 per cent of GNP in 1963 up to almost 9 per cent in
1980).
Especially structural and fundamental legislation can be seen as
an expression of this development. These laws concern the financing and planning of health services. In 1982, Parliament passed
the Health Services Act which replaces the Hospital Facilities
Act 1972. Both Acts were designed to achieve a (better) regulation with respect to the volume and to the organisation of the
health services. The following description of both acts is taken
from Rutten and Van der Werff (1982).
From the date of enforcement of the 1972 Act it was no longer
allowed to build or extend a hospital facility which required
an investment of more than 500.000 of Dutch guilders without
a license granted by the Minister of Health. From 1979 onwards
the Hospital Facilities Act was replaced by the Amended Hospital
Facilities Act. This act was designed in a more flexible way
and opened up the possibility of drawing up regional hospital
plans for categories of facilities on the basis of an instruc-
tion of the Ministry of Health. These plans developed by the
Provincial Governments needed the approval of the Minister of

Health. The Council of Hospital Facilities advised in this
respect. On the basis of the approved plans licenses were
granted for constructing or reconstructing hospital facilities.
A new element introduced by the Amended Hospital Facilities Act
was the possibility of "closing down" facilities.

The Health Services Act 1982 will cover regulation of the
whole of the health services including both institutions and
privately practising professionals. It will probably be implemented in 1983. It introduces decentralization of administrative
responsibilities to the level of provinces and municipalities.
This new legislation also comprises regulation of the participation of patients and clients in the decision-making process.
Finally, developments in financial legislation and measures
should be mentioned. Control of hospital costs was regulated by
the Hospital Tariffs Act. Recently this Act was replaced by
the Health Tariffs Act, covering both institutions and individually practising professionals. The previously mentioned
development to finance ambulatory services from the General
Special Sickness Expenses Fund, can be seen as a measure to
stimulate coherence of the financial system.
In connection with these developments in our health legislation,
we want to make the following remarks. The new legislation
could be a useful instrument for stimulating cohesion and
coherence on our health services system, if it could contribute
to two developments:

a. Enlarging the decomposability of the health services system,
   by means of regionalization. This means that a certain amount
   of the existing relationships in the health services system
   are to be diminished or weakened, for instance most of the
   relations between regions.
b. Enlarging the coherence of the control system and the financial system, in other words, the interdependency between aspects which have to be controlled together must be strengthened.

Nevertheless we wonder whether this legislation and related
measures will be able to make for such a contribution. In the
next part we will go into this subject.

3.2. Dangers and counterpressures

Although the legislation mentioned above is aiming at better
controllability of the health services system, there is a lack of
a coherent control system or decision-making system on every level
of health services at this moment.
This can lead to the following assumptions:

a. the existence of unnecessary relations which lower decomposability (and thus controllability) of the health services
   system;
b. the lack of relations which should have existed to offer a
   coherent health services system to clients;
c. the non-correspondence of the governing bodies and the structure of the health services system; this means that there is no
   correspondence between relatively autonomous part systems and
   the governing bodies;
d. insufficiently guaranteed coordination between these governing
   bodies (because of the weak interactions between the partsystems);
e. the non-correspondence of the controlling system in practice
   and the theoretical notions of hierarchical systems which were
   presented in part 1 of this paper. Especially as far as it concerns directives for planning: according to these notions they
   are far too detailed.

      The lack of such a coherent system leads to the phenomenon
that decision making processes move in a central direction
(Leenen, 1979).
In so far as our analysis may be proven right, we can conclude that
the effect is counterproductive.

      In considering the enforcement of the Hospital Facilities
Act, these developments can be noticed clearly.
It is true that this Act allocated planning tasks to the Provincial
Government; but it included such an overwhelming amount of
detailed directives concerning the desired ways of planning that
in fact there was almost no room for decentralized planning and
decision making.
The Health Services Act offers more room for decentralization,
but no guarantees are built in to prevent the Central Government's
detailed involvement in planning (De Leeuw, 1982).
Moreover it is noteworthy that the causes of this tendency to centralization are to be found not only at the State Government.
Leenen (1979) points out that the private institutions and health
care workers could exert influence on the control of the health
services system, if they gave up their autonomy aspirations. It
is true that cooperation and selfordering do not necessarily lead
to the abolition of legislation, but they diminished the area
of legislation. However, selfordering nearly did not develop in
the Dutch health services system. The tremendous commotion in the
private sector over the detailed norms in the Hospital Facilities
Act can be seen as a very remarkable phenomenon in this context.
The objections mainly concerned the size of the norms and insisted
on refinements and more detailed norms instead of the rejection of
so many specifications.
Moreover it is important to notice that processes like centralization and decentralization are always partly a matter of

power aspirations and self-interest (Leenen, 1979). In both
cases there is always someone who loses power. Often the delusion
is made that centralization leads to an increase of power for
the central bodies and the other way round. It must be noticed,
however, that losing power (in the sense of giving more room for
decision making) to the lower governing bodies can lead to increasing power (in the sense of real influence).
This curious phenomenon is caused by the fact that centralization
can lead to the following consequences:
- the controllability of the whole can decrease
- the controlling capacities of the governing body can fail
- the controlling capacities of the decentralized governing bodies
  will not be used anymore
- the decentralized controllers will pay attention to the control
  of the central governing body instead of the control of their
  own institutions.

    Bureaucratization in the negative sense of the word can be
considered as one of the most serious consequences of the centralization process.
This phenomenon leads to the emergence of heavy, rigid governing
bodies, which tend to pay much more attention to themselves than
to the health care system; in other words, most or all attention
is being paid to the control of the governing bodies, the preparation and enforcement of their legislation, regulation and
measures.
This results in a tremendous excitement, e.g. many memoranda,
reports and meeting, by which the originally intended effect,
i.e. a coherent health services system, controlled by a coherent
control-system, has disappeared. Sight is lost of the main objective and the subjects in the health services system, namely
health care and the participants in health care.
Government planning is developing now without much consultation of
the private institutions and workers. This happens while it is
reasonable to assume that policy making, directed to change, calls
especially for intensive communication with those who have to
effect the change!
In this context Leenen (1979) points out that not reacting to
advices and making decisions without consultation of those concerned, are guaranteed means to lose goodwill, which is necessary
to effect real changes. In view of the foregoing analysis it can
be assumed that the tendency to centralization and the lack of a
coherent decision making system, which can be noticed in the Dutch
health services system, will both interfere with the emergence of
coherence.
Leenen (1979) mentions other factors which are also stimulating
this lack of coherence, namely:

- a variety of health services institutions
- growing specialization

- stereotyped ideas and attitudes; sharpened roleconceptions and
  increased status differentiation between professionals
  between institutions
- differences in growth between the various sectors and echelons.
Another factor can be added, namely
- the unintegrated financial structure, as described in part two.

These factors can be seen as an obstacle, the more so as the
connected control-problem is not well defined. For there are two
interconnected control-problems.
a. The question what implies a well functioning and controllable
   health services system.
b. The question how changes in the direction of a better situation
   could be achieved.

The second question (concerning the process of change) is
virtually not under discussion. Structural and cultural aspects
of the control process simply have to bridge the gap between
structural and cultural aspects of the existing system on the one
hand and the same aspects of the future-desired system on the
other hand.
Sometimes the impression is given that the existing system is controlled as if the future was already there.

4. Possibilities to stimulate well-considered cohesion and coherence

Despite all obstacles against the aspirations to reach more
controllability and coherence in the Dutch health services system,
it is necessary to keep on looking for possibilities and means to
stimulate this.
Enlarging cohesion and coherence frequently is considered as an
important objective, which could be achieved by building up all
kinds of new relationships.
We suggest that this idea is based on a delusion. Breaking up
relationships, which are diminishing controllability has to be
preferred to building up all kinds of relationships. For the
system has to be composed of relatively autonomous partsystems.
Following this assumption we can conclude that it is important
to create well-considered relationships instead of all kinds of
interrelations between partsystems at any cost.
To reach such a cohesion and coherence the following measures
can be mentioned.

a. Decentralization

In this context we think of real aspirations to and
enforcement of decentralization.
This means that, in the frame of a closely-structured system

the discerned functions are situated at different control-
levels and then coupled, by assigning responsibilities to
the right levels (Leenen, 1979). In general it can be
assumed that maximum decentralization is needed, but there
must not be more decentralization than is called for by
coordination-requirements.
Put in another way, coordination must not be executed on a
"higher" level in the control-system than is strictly needed.
This i.a. can be achieved by regionalization. In our opinion
two aspects are covered by the concept of regionalization,
namely
- decentralization of existing and newly created decision-
  making functions to regional health services systems in
  the frame of the closely-structured total-system (Ministry
  of Health and Environmental Protection, 1974);
- accomplishment of a stable and closely-structured health care
  system in every region. These regions will have to function
  as relatively autonomous partsystems. Moreover, this idea
  must be applied to other hierarchical levels: districts,
  towns, quarters of a town and so on.

   Regional planning, coupled to regional budgetting can be
considered as mechanisms to stimulate regionalization.
These ideas are not new. The case for regional planning together with regional budgetting can be heard more and more
in The Netherlands.
Civil servants and researchers as well as managers in health
services institutions have been asking for this during recent
years.
Already in the "Memorandum on the Structure of Health Services"
(Ministry of Health, 1974) the proposal was made for regional
budgetting. More recently - in May 1982 - these ideas were
under discussion at a conference of the Interacademic Institute of Health Sciences (I.W.Z.). Some scientists put forward
objections to such a system of regional budgets. For instance
Ter Heyde and Kalff (1976) expected regional budgets to weaken
cost-control: neglected regions could emerge, which would
aspire to the level of the best provided regions.
Moreover this system would stimulate inequalities: in some
regions people would have access to better facilities than in
other regions. We encounter here a fundamental dilemma, namely
the dilemma of justice on the one hand and controllability on
the other hand; a viable complex system has to be composed of
relatively autonomous part systems, which are composed of relatively autonomous part systems and so on.
For the structure stimulates stability and controllability.
However, if facilities are to be distributed in a just and
defensible way, it is necessary to distribute them centrally and
uniformly, either directly or indirectly, by way of standardization. The more a uniform distribution is practised, the more

the controllability and stability is destroyed. In the
worst case this kind of distribution leads to the situation
that no one will receive the care he needs. The principle
of justice, considered in the sense of "what is a sauce for
the goose is a sauce for the gander", in the extreme case can
lead to the situation that there is no sauce anymore, not for
the goose nor for the gander.
Moreover, the assumption can be made that uniformity for the
sake of justice often means no more than a would-be justice.
For instance, standardization based on density of population
is not justified when it is applied in sparsely populated
areas. In these areas facilities are not in easy reach for the
inhabitants. The objective can be met by preseribing a basic
package of health care services in every region and simply accepting a variation in facility-patterns by region.
Despite the objections to regional budgeting and despite the
problems accompanying the enforcement of this system, we consider regional budgeting as a requirement, necessary to achieve
useful regional planning. Regional budgeting also means that
budget-holders will have room enough for planning, budgeting
and distribution of finances.
For if those functions are not assigned to the regional level,
it will be impossible to realize regional plans.
In our country tendencies can be noted towards this regional
planning and budgeting. The health legislation mentioned before
is one example and the (plans for) experiments in pilot-regions,
coupled to (scientific) research and evaluation is another.
For the rest it should be necessary to guard against endless
experimenting without real changes in the system being achieved.
In that case experiments will be nothing else but objections to
aspirations to more cohesion and coherence in our health services
system. Other counter-pressures have been under discussion
before, like the tendency of State Government to centralization.
As noted before, this tendency can be observed in the detailed
directives associated with the allocation of decentralized
planning tasks.
So the State Government should substitute blueprint-planning by
a system of so-called frame-planning; in this case frames are
indicated by broad directives, which leave room enough to the
lower authorities to make their own plans.
Moreover, a change of attitudes and working-methods is needed
within institutional health care. The regional cooperating
bodies, in which the institutions are participating, will have
to be more than mere consultative bodies. Certain functions must
be assigned to these bodies. However, institutional managers
often do not realise that, despite - or may be thanks to - the
assignment of functions to regional bodies, their own influence
on decisions, which are made elsewhere at this moment, could be
enlarged (Leenen, 1979). It is true that managers of these institutions will give up then a small part of their own indivi-

dual autonomy, but in return they will get a more autonomous position collectively with regard to the public authorities (Mur-Veeman, 1981). The public authorities will not be
in the position anymore to play off the institutions against
each other.

b. Restructuring the financial system

Restructuring our dispersed financial system can be
mentioned as another possibility to stimulate cohesion and
coherence in our health services system. The Health Tariffs
Act, which has come into force from 1982 onwards, can be
considered as one step in this direction, just like measures
directed to more uniform financing of ambulatory facilities.
To stream-line the financial system, however, can bring the
danger of centralization; so much more a reason to introduce
regional planning and regional budgeting and may be impose
regional premiums. Schrijvers (1980) has worked on this matter.
In this context he proposed three possible models. In the first
place he mentioned the so-called English model. According to
this model the regions would receive their finances from one
single fund. This would create the possibility of financing on
the basis of relative needs. In this case there must be only
one single general insurance-system. A second possibility is,
according to Schrijvers, the so-called Swedish model. When
using this model regional authorities would receive their
finances from different sources and thereupon would make
finances available to the health care service bodies.
In the third model institutional budgets and tariffs would be
fixed at a national level. Regional authorities would advise on
the extent that institutional budgets fit into operational
and strategic plans.
Schrijvers prefers the third model, because it would not disturb
the existing flows of money.
We prefer the Swedish model, since it would counteract the tendency to centralization, but not lead to such a fundamental
change as required in the English model.
Moreover, the English model does not fit very well in the specific Dutch situation, characterised by religious private initiatives, which can be seen as essential elements of our culture.

c. Restructuring the educational system and educational programs

In our opinion the reconstruction of our educational system
must refer to at least two aspects, a substantive aspect on the
one hand and a control aspect on the other hand.
The substantive aspect concerns the level and nature of knowledge in educational programs for civil servants and health
care workers. Final educational requirements have to be revised
in order to reach a closer correspondence to professional and

practical requirements, especially in view of the present
process of change to more cohesion and coherence in the
health services system.
In this connection we think of knowledge to enlarge skills and
expertise concerning control, management and organization on the
one hand and knowledge to change stereotyped ideas and attitudes, like sharpened roleconceptions, status-feelings and status-
differences, aspirations to autonomy and resistance to change
on the other hand.
In the educational programs of medicine, for instance, more
knowledge must be offered about problems of organizations and
policy-making. This could enable future doctors to contribute
to the solution of this kind of problems, if necessary. Besides,
a better insight into organizational problems may result in the
situation that doctors will realise better the organizational
consequences and costs of their activities, plans and wishes
(Mur-Veeman, 1981). The medical students should also be taught
to assess their own knowledge in the right way and, if
necessary, to realise the relativity of this knowledge.
This also applies to other professions in health care.
Moreover, future civil servants and future controllers and
managers in health care should be enabled to obtain knowledge
about decision-making processes, planning methods, cooperatio-
nal relationships, control and organizational problems and so
on.
We may note that at present such an educational program is being
developed at the State University of Limburg.
Improving and enlarging possibilities of additional schooling,
especially for managers in health care and civil servants, is
also to be recommended. The second aspect of restructuring the
educational system, the control aspect, concerns the development of policy measures to create possibilities of changing
the existing educational programs or developing new educational
programs. Besides,development of mechanisms to regulate input
and output of students can be mentioned in this connection.
These measures are suggested to achieve a better correspondence
between needs and requirements of the health services system
on the one hand and final requirements and results of the educational programs on the other hand.

d. Continuity of political decision-making

It can be noted that political office-holders enjoy only
relatively short terms of office. A structural process of
change, as described above, will require a relatively long
period of maybe 15 to 20 years.
It is understandable that politicians aspire to achieve some
concrete successes during their four year term of office.
This phenomenon sometimes leads to the achievement of seemingly
short-term successes (like the reduction of the amount of

beds), which might not contribute to essential structural
changes on the long term.
Therefore it is recommended to examine seriously how longterm
processes of change can be controlled, despite change in political leadership.

5. Suggestions for research

On the basis of the previous analysis of the Dutch health
services system the following research themes are suggested:
- research into the functioning of the whole, analysing control
  and controllability;
- research into educational programs;
- analysis of power aspirations and power mechanisms in the health
  services system;
- functioning of political decision-making.

Summary

The health services system in The Netherlands can be characterised by its overwhelming complexity.
The following problems, noted about this system, can be seen as
closely associated with this complexity:
- the system is considered to be incoherent
- uncontrollability is one of the most outstanding characteristics of the system
- the system is considered to be too costly
In this paper we described the structural aspects of the system and
its lack of coherence in terms of systems- and controltheory.
Then the measures being proposed and taken by the State Government
in order to tackle the problems were analysed. In this context the
structural and financial legislation were described.
There are, however, strong resistancies and mechanisms, which might
result in decreasing coherence and controllability, like tendencies
to centralization, misconceptions about the appropriateness of
policy measures and about power regulations and power mechanisms,
and resistance to change.
Thereupon some measures were presented which may provide positive
results: real decentralization, restructuring the financing system
and changes in the educational system.
Finally some research themes were suggested, including the functioning of the whole system, educational programs and power mechanisms.

References

Conant, R.C., "Information flows in hierarchical systems" in:
International Journal of General Systems, Vol. 1 (1974) 1.

Conant, R.C., "Laws of information which govern systems" in:

I.i.e.e. transactions on systems, men and cybernetics, vol. 6
(1976) 4.

Field, M., "The concept of health systems at the macro-sociologi-
cal level" in: Social Science of Medicine, (1973) 10.

Heide, H. ter and D.J.A. Kalff, "De macro-organisatiestructuur
als hulpmiddel bij de beheersbaarheid van de kwaliteit in de
gezondheidszorg" in: Beheersbaarheid van kwaliteit, kwantiteit
en kosten, Lochem: De Tijdstroom, 1976.

Koestler, A., Janis: a summing up, New York: Vintage books, 1979.

Leenen, H.J.J., Structuur en functioneren van de gezondheidszorg,
Alphen aan den Rijn/Brussel: Samsom Uitgeverij, 1979.

Leeuw, A.C.J. de, "Organizations, systems, wholes and parts.
Variations on a theme in organizational science",in: Methodology
and Science, Vol. 13 (1980) 4.

Leeuw, A.C.J. de, "Men gaat niet de gezondheidszorg maar een stel-
sel van richtlijnen besturen", in: Het Ziekenhuis, (1982) 1.

Ministerie van Volksgezondheid en Milieuhygiëne, Financieel Over-
zicht Gezondheidszorg (Financial Survey), 's-Gravenhage: Staatsuitgeverij, 1981.

Ministerie van Volksgezondheid en Milieuhygiëne, Structuur-
nota Gezondheidszorg (Memorandum on the Structure of Health
Services), 's-Gravenhage: Staatsuitgeverij, 1974.

Ministerie van Volksgezondheid en Milieuhygiëne, Richtlijnen
ex artikel 3 van de Wet Ziekenhuisvoorzieningen, 's-Gravenhage:
Staatsuitgeverij, 1981.

Mur-Veeman, I.M., Ziekenhuisbeleid, Oss: Witsiers, 1981.

Rutten, F. and Werff, A. van der, "Health Policy in the Netherlands
at the Crossroads", Maastricht, 1982.

Schrijvers, G., Regionalisatie en financiering van de Engelse,
Zweedse en Nederlandse gezondheidszorg, Lochem-Poperinge:
De Tijdstroom, 1980.

Simon, H.A., "The architecture of complexity" in: Proceedings of
the American Society, 106 (1962) pp. 467-482.

3.1.2.6. Transactions on Systems, Man, and Cybernetics

THE ITALIAN HEALTH SERVICES SYSTEM

A. Bariletti

Institute of Science and Finance, University of Rome

Rome, Italy

Foreword

Italy is in the course of reforming her health services system. In this phase of transition, administrative and financial aspects appear to be crucial for any future development. Accordingly, particular emphasis will be placed on these issues throughout the following discussion.

0.1. Values and Objectives underlying the Health Services System

The Constitution of the Italian Republic acknowledges health care as a basic right to which the individual members of the collectivity are entitled, as well as a fundamental issue of public interest.
With the Health Services Reform Act (1978) the Italian Parliament has approved the foundation of the Servizio Sanitario Nazionale (S.S.N: National Health Service), therefore dismissing the previous system, fragmented in various health insurance funds managed by different and separated public institutions.
Thus, the main objective of the promotion and protection of the health of the individuals within the nation, has been assumed among the direct functions and responsibilities of the State, to be pursued through the newly founded S.S.N. under conditions of public financing and equality of access.
The Health Services Reform Act (1978), as the normative framework within which the new system should be developed, seems to be inspired by the following general principles:

- unitary approach to the organisation of the specific services and functions
- comprehensiveness in the organisational action
- universality of beneficiaries.

According to the Health Services Reform Act (1978),the S.S.N.,
as a 'set of functions',should aim at the following targets':
- to educate individuals as to develop a relevant degree of
  awareness towards the health care issue;
- to prevent diseases and work injuries;
- to reduce mortality, morbidity and disablement through
  diagnosis,cure and rehabilitation;
- to promote and guarantee healthful living and working conditions;
- to strengthen the degree of territorial equity in the production and consumption of 'better health' (i.e. to remove
  interregional and intraregional disparities);
- to promote responsible motherhood and healthful childhood;
- to provide health care, as well as to prevent and eliminate
  any form of discrimination and deprivation, for old-age
  and mentally handicapped people;
- to enhance primary health care and to integrate health and
  social services;
- to control and remove environmental pollution and other
  forms of hygienic and sanitary hazards.

0.2. Basic Political Configuration of the Nation

The Parliament of the Italian Republic is structured
according to the principles of a bi-cameral system. Besides
the legislative power, the Parliament exercises political
control over the Government in charge; approves the Financial
Act (Budget) and other financial bills; etc.
The general organ of the Government is the Cabinet. The Head·
of the Cabinet is appointed by the President of the Republic.
Ministers are selected by the Premier, who is in charge of
directing and coordinating the Cabinet's activities, but are
appointed by the President of the Republic.
Government administration is structured in various Ministries
which are diversified by function; at the local level central
Government's action is exercised through branches of the
different Ministries.
From 1947 political power has been predominantly exercised by
multi-party coalition Governments. Political parties in
Italy are numerous (1979 : 9); as a consequence setting up
coalitions may be difficult, and this may partially account
for the proverbially high mortality rate of the Governments
in charge, as well as for particularly long legislative
(technical) times.
At present, the coalition is formed by five political parties

(Christian Democrats, Liberals, Republicans, Socialdemocrates, Socialists) The Ministry of Health has been in the past legislature, and still is, allotted to the Liberals.

The Italian Constitution prescribes a democratic organization of the State, decentralized at the territorial level in Comuni (Municipalities), Provincie (Provinces), Regioni (Regions).
In such a structure, the Comuni are the elementary units; the Provincie are sets of Comuni, and the Regioni are 20 autonomous public bodies, with legislative powers within the limits set forth by the Constitution.
Five Regions have been granted special statutes and, accordingly, greater administrative and legislative powers, because of particular ethnical, geographical and economic conditions.

1.1. The Pre-reform System of Health Services

In trying to outline and evaluate the actual profile of the Italian health services system, one has to bear in mind that the system has entered in recent years (from 1977 onwards) a transitory phase of structural modifications, which is at present in its very beginning.
As mentioned before, the system is moving from an extremely heterogeneous and fragmented structure formed by a multitude of Funds for Compulsory Health Insurance, towards the 'global' S.S.N. framework, envisaged by the Health Services Reform Act (1978).

1.1.1. The General Structure of the Pre-reform System

In the system previously in force, fully developed between the forties and the seventies, health care was provided by public as well as private institutions.
Public care was provided by the State in specific cases of infectious and socially relevant diseases (e.g. V.D.; poliomyelitis). Provinces provided psychiatric care and Municipalities provided general health care, to indigent and needy people, as forms of public welfare relief of charitable origin. The bulk of health care (92% of the total population coverage, in 1971) was provided by many different compulsory sickness insurance funds.
In 1971, INAM (National Institute for Sickness Insurance), ENPAS (National Provident Fund for State Employees), ENPDEP (National Provident Fund for Public Corporations Employees), INADEL (National Insurance Fund for Employees of Local Authorities), Cassa Mutua Artigiani (Mutual Benefit Fund for Self-employed Farmers), Cassa Mutua Esercenti Commerciali (National Benefit Fund For Tradesmen), together covered about 88% of the population.

The remaining 4% was covered by some 200 other Funds of
various origin (e.g. local authorities; professional
groups; private firms; etc.) and dimension, with differing
contribution and benefit rates.
The following discussion will mainly refer to INAM, as the
most significant national compulsory insurance fund.

1.1.2. Operational Characteristics of the Pre-reform System

By far the largest Fund (53% of the total insured population, in 1971), INAM was a public corporation (regulated by
the Treasury and the Ministry of Labour and Health).
Administered at the local level through 25 provincial
branches, INAM provided medical care coverage, maternity insurance and sickness insurance benefits, to agricultural, industrial, commercial, banking and insurance workers in the private sector of the economy, as well as medical care coverage
for old-age pensioners and recipients of incapacity pensions.
Funds were raised through contribution charges paid for the
major share by employers and for the remaining portion by
employees. Rates however varied widely between insured members, according to different branches and sectors of activity and to occupational level and status.
As far as medical care was concerned, INAM provided in-kind
benefits for general practice, specialist services, hospital
care and pharmaceutical products. Doctors and hospitals
entered into agreement with INAM and were paid directly by
the Fund on fixed rates. Usually doctors' remuneration
systems could be per item (specialists), per capita (GP's)
or a mix of the two; hospital charges were dependent on the
type of treatment.
Qualifications for access to all types of treatment was based
on occupational status; patients registered in a list with
the physician(GP's) of their choice and were granted access
to further treatments (hospital services, specialistic care)
 through a 'certificate of need' issued by the(GP's.)
Prescribed treatments and pharmaceutical drugs were free of
charge on the point of service, with the exception of dental
care.
INAM medical care benefits were granted for a maximum period
of 180 days per year; further benefits should have been met
out of personal purchases.
In special cases Regional and Provincial authorities could
offer additional coverage (e.g. rehabilitation cases). Free
treatment could be obtained by Public Welfare at the local
level, provided the individual registered in the poor list.
Other public sector Funds had roughly similar coverage level
as INAM, while benefit provision was mainly centered on

hospital and specialistic treatment, with little or no
coverage for drugs and GP's. services, in professional group
and selfemployed insurance schemes.
Sickness benefits were paid by INAM to insured members for
incapacitation by illness and t.b.. Benefit rates varied
between 1/2 and 2/3 of the daily wage rate and were limited to
180 days, with no allowance for the first 3 days.
In special cases supplementary benefits were granted by the
Fund to alleviate conditions of particular hardship. A 50%
contribution could normally be obtained to supplement
prothesis purchases by insured members.
INAM also provided death allowances and maternity insurance,
the latter in the form of cash benefits (as a lump sum or
a percentage of the daily wage rate) and access to surgery.
Other public Funds did not provide equal sickness benefits as
INAM, as these were frequently accounted for in the collective wage bargain.
Besides being supplied by insurance Funds, medical care was
obviously purchased by individuals through the private market,
as no limit but ability to pay was set to private spending.
In 1971, about 15% of total health expenditure was met out
of private spending; the remaining 85% was financed by the
Funds (52%) and by the Central and Local authorities (33%),
the latter to make up for the Funds' chronic deficits.
The preceding rudimentary discussion could not do justice to
the extreme complexity of the pre-reform health service
system structure. However it should enable one to grasp the
notion that the system was characterized by an almost
exasperated degree of heterogeneity as far as conditions for
qualification, extension and types of coverage, contributory
rates and benefit ranges were concerned, with little engage-
ment for basic preventive action.
On the whole the pre-reform system structure could be defined
as the mixed outcome of various historical, economic and
social forces : reflecting the different degree of political
power exercised by the relevant pressure groups in legiti-
mating their access to the health care sector, as well as
fragments of redistribution policies enacted by the public
power systems.
This system has been put into liquidation with the 1977
reform act, and is now beginning to get reorganized according to the S.S.N. scheme envisaged by the 1978 act (cfr.
infra, 1.3, 1.4).
Although several technical characteristics of the previous
system are still in operation (e.g. doctors' remuneration
systems, individuals' contributory mechanism, contracted
agreements between public authorities and private hospitals,
etc.), health care should be provided by the newly founded
S.S.N. in a way consistent with its innovative principles.
Fundamentally, discrimination in health care coverage due to

different occupational status should be removed and preventive
action enhanced.
Sickness benefits, previously issued by the Funds, will be
maintained, as their administration has been transferred to
INPS (National Institute for Social Insurance), the public
corporation already in charge of superannuation benefits.

1.2. Related Forms of Social Protection; Public Welfare
Services; Private Insurance Schemes

     Of related interest to health care coverage, are the
benefits provided through:
a) the old-age and disability pension scheme;
b) insurance against work injuries;
c) total and partial unemployment insurance;
d) family allowances;
e) income tax deductions for health care expenditures;
f) income tax deductions for family allowances;
g) public Welfare services.
INPS actually manages the major share of a), c) and d);
INAIL (National Institute for Insurance against Work Injuries)
administers b).
Personal income tax deductions for health care purchased in
the market have been recently (1981) introduced and are
allowed according to a flat-rate scheme or, alternatively,
to effective expenditures up to a fixed ceiling.

     Public welfare administration is also extremely complex
and heterogeneous. At a first level of intervention, "institutional" welfare assistance to needy families is provided in
kind, by Regional authorities. Supplementary assistance is
then issued by the State, the Municipalities and the Provinces (e.g. psychiatric care; aid to incapacitated people;
assistance to deaf and blind; etc.).
Besides,various publicly funded institutions (mainly originated on a professional-group basis) of differing size and
scope provide specific types of assistance and relief to
particular groups (e.g. workers' orphans; incapacitated
workers; etc.).
Private and religious charities are also very active in this
sector. If public welfare is considered on the whole, it is
worth noting that Government action succeeded, during the
seventies, in unifying main welfare activities and transferring
them to the Regional level. However, the existence of the
above mentioned institutions is causing criticism from an
economic and political point of view because of their very
dubious degree of effectiveness.
The present public welfare system is expected to be reformed
in the future, to be integrated in the S.S.N.

1.3. The Reformed Health Services System Structure

1.3.1. The S.S.N. General Scheme

According to the Health Services Reform Act (1978) the
reformed system as represented by the S.S.N. should be
decentralized and organized on three levels of government,
as follows.
A. Central level:
   1) the Parliament establishes the general legislative
      principles to which decentralized actions should be
      informed and approves the Piano Sanitario Nazionale
      (P.N.S. : national health plan);
   2) the Government together with the Ministry of Health
      formulate the P.S.N.; control resource allocation;
      direct and coordinate decentralized activities; promote
      research and development;
   3) in framing its action, the Government is assisted by
      the Consiglio Sanitario Nazionale (C.S.N. : national
      health council), appointed to be consulted on health
      policy matters.
B. Regional level:
   the Regional authorities are in charge of the following
   activities:
   1) to found the U.S.L.'s (Unità Sanitarie Locali: Local
      health services units) as a comprehensive set of
      health services structures and functions operating over
      a specified regional area;
   2) to regulate and control the U.S.L.'s from the point of
      view of efficiency and effectiveness;
   3) to formulate the Regional health plans;
   4) to organize and promote health manpower policies with
      special reference to professional training for nurses
      and ancillary personnel.
C. Local level:
   1) Municipalities are in charge of running the administra-
      tion of the U.S.L.;
   2) the U.S.L., envisaged as the basic decentralized organ
      of the S.S.N., is in charge of promoting and providing
      comprehensive health (and social) services through the
      establishment of the Distretti Sanitari (D.S. : Sani-
      tary Districts) and other elementary units (e.g. hos-
      pitals; surgeries ; etc.) within the relevant geographi-
      cal area;
   3) the D.S.'s are the primary units in charge of health
      care privision at the distributive local level.
      Until recently, private health insurance schemes were
      practically non existent in Italy. It is interesting
      to remark that these schemes were firstly introduced

on the market almost contemporaneously with the
reform action pursued by the Government in the health
care sector.
in 1979 all the 32 authorized insurance companies marketed their policies. These were meant to supplement
public care provision in the transitory phase between
the dismissal of the pre-reform Funds and the full
establishment of the S.S.N. The schemes were directed
especially to middle-class self-employed professionals,
typically making available: more choice of treatments,
more comforts, some cash benefits and less time costs.
The health market share attributable to insurance companies is still minimal: in 1980 insurance policies
revenue could be roughly estimated at about 1/1000 of
global public health expenditures.
At present it is not clear whether these policies (due
to their restricted range of benefit coverage and
their limitation to curative treatments only) could
also be envisaged as pure substitute of publicly
provided health care, in the long run.
It is interesting to note here, that although insured
members of the pre-reform Funds are automatically
eligible for coverage under the S.S.N., no lawful
obligation has been set to join the public system and
"opting out" seems in principle admissible.

1.3.2. The P.S.N. Planning Framework

The P.S.N. is envisaged as the basic tool of indicative
planning at the national level.
In the P.S.N. (valid over a period of three years) are
framed:
a) the strategic aims and the specific target-programs and
   sub-programs towards which local decentralized action
   should be oriented;
b) the amount of the Fondo Sanitario Nazionale (F.S.N. :
   national health services financing fund) to be annually
   inserted in the Budget (Financial Act);
c) the standards to allocate F.S.N.'s quotas between the
   Regions;
d) the guidelines to share quotas between the U.S.L.'s at
   the local level;
e) the general norms and standards for health care provision
   and manpower training;
f) the general rules to be accomplished for monitoring the
   P.S.N. feasibility and implementation through the establishment of S.I.S. (servizio informativo sanitario:
   health services information system).

1.3.3. The S.S.N. Financial/Administrative Circuit

A better grasp of the planning mechanism and coordination between different administrative levels should be obtained examining fig. 1, that synthesizes the financial/administrative circuit of the reformed health services system.

It may be useful to note that in phase (A), the F.S.N. is earmarked in the economic forecasts of the Treasury(consumption expenditures) and of the Ministry of Finance (capital expenditures). The F.S.N. is eventually appointed in the Financial Act, to be approved within the Budget.

In phase (B) the proposed fund sharing between Regions should be approved by the C.I.P.E. (Interministry Board for Economic Planning); in phase (C) U.S.L.'s should manage their economic and financial budgetary schemes according to uniform codification.

Phase (D) is of particular relevance as it refers to the financial and administrative management of the effective supply units at the local level. From a formal point of view coordination at the U.S.L. level can be decomposed as follows:

- Municipalities are expected to plan yearly U.S.L.'s budgets and programs, and to exercise bi-monthly controls on U.S.L. expenditures to detect deficit forming.
- The Regional authority (supervising the various U.S.L.'s in a certain area) is expected to relate to each U.S.L.'s management trend and financial situation, comparing stated programs with effective expenditures.
- The U.S.L. should provide, via the Region, the Ministry of Health and the Treasury with quarterly data on forecasted balances.
- Banking institutions, acting as U.S.L.'s treasurers, should provide similar data on effective cash-flows.
  In case of deficit forming (unrelated to exceptional morbidity or mortality conditions), U.S.L.'s are held responsible of excess expenditures and Municipalities are expected to adopt measures to balance the budget.
  (Comments on the S.S.N. administrative circuit, mainly connected with financial and decision-making problems, will be presented sub. 3 and 5.).

1.4. Health Care and Social Services

Social services provision should be enhanced at the local level, as U.S.L. are expected to integrate different forms of social protection, adapting to the target-programs suggested in the P.S.N. Comprehensive action will, however, face constraints due to the still heterogeneous public welfare system.

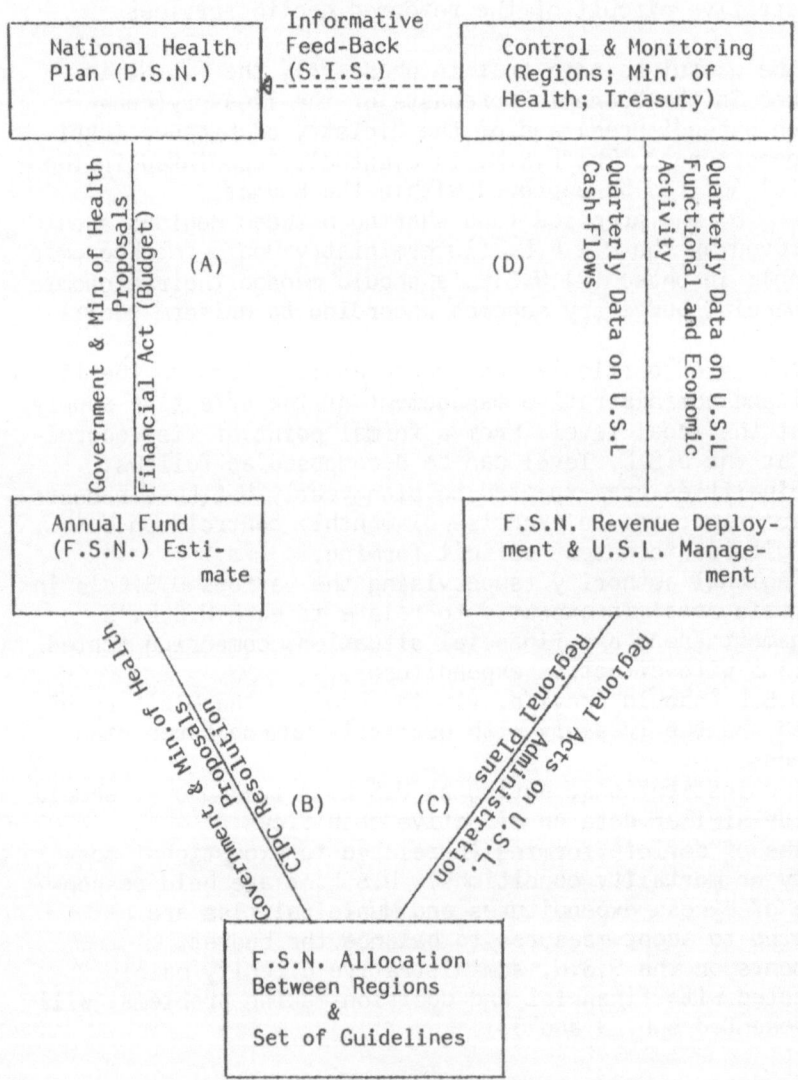

Figure 1   The S.S.N. Financial/Administration Circuit.   Source:
         Ministero della Sanita.

2.1. Problems in Manpower Policies

Manpower distribution as inherited from the pre-reform system is unsatisfactory and characterized by an excess of physicians in comparison to ancillary personnel, as well as by regional disparities in staff availability. While the latter problem is fairly common in the majority of health systems, the former is typical, in that Italy has regularly exhibited a high (and until a few years ago, steadily increasing) doctor/population ratio (1975: 1/20.000) together with a low quota (1975: 30%) of professional nursing personnel over the total of health services ancillary personnel. The main causes that accounted for this serious unbalance have been:
i) the absolute lack of any coordination between the educational system and the health system (e.g. students' access to Medical Faculties was unrestricted); ii) the lack, in the pre-reform system, of any planning mechanism apt to implement manpower policies from the point of view of the system as a whole.
According to the Health Services Reform, Regional authorities are expected to activate manpower training for ancillary personnel, while Governmental ability to regulate and restrict access to Medical education will prove crucial to redress the balance in physicians' availability. Measures have already been taken in this direction, to regulate Dentistry schools.

2.2. Health Expenditure Trend

Table 1. provides a breakdown of health expenditures by type, as percentage of total, in 1977 and 1981.

Table 1. Expenditures by type, as percentage of total, 1977 and 1981.

	1977	1981
Preventive treatments	4.9	4.8
Health Services Institutions	37.4	34.6
Public Hospitals	49.0	50.3
Regional Expenditures	0.7	0.6
Contract Hospitals	8.0	9.7

Source: Ministero della Sanità

It is evident that the expenditure trend has not yet been
modified within the newly founded system: hospital and
personal health services are still the most relevant
resources-consuming items.
To reverse this tendency will prove to be a major task for
the S.S.N. : the present health care structure is rigid and
biased towards curative outcomes, while financial difficulties (cfr. 3) and effective decision-making and allocative
patterns (cfr. 5) seem to constrain attempts to enlarge the
system's range of activities, at least in the short run.

3.1. The Growth of Health Expenditures

Table 2. gives estimates of health care expenditures as
percentage of G.N.P in various years.

Table 2. Health care expenditures as % of G.N.P, various years

1975	5.80
1976	5.71
1977	5.66
1978	5.78
1979	5.99

Source : Ministero della Sanità

S.S.N. expenditures have been estimated at 5.8% of G.N.P.,
in 1982. It has already been noted that these expenditures
consist of pure transfer payments from the State to the
Regions, as these are not entitled to dispose directly of
funds raised through the contributory system, nor do they
possess autonomous taxing power in this respect.
(It should be added that the old contributory system has
been only slightly modified: rates have been raised and
employers' contributions have been detaxed. Individuals
previously not insured with the Funds are entitled to
coverage under the S.S.N., provided that they contribute a
lump-sum yearly charge).
At present (1982) health expenditures transfers amount to
13-14% of the total payments form the State to the local
authorities and are, in this respect, the largest single
expenditure item in the national budget.
Apart from the absolute figures, the most disturbing aspect
of the present financial situation is the almost automatic
mechanism through which these expenditures are generated.

In theory and according to the financial/administrative circuit previously described, the Government should determine
ex-ante the F.S.N. annual amount and allocate it between the
Regions.
In practice however, this procedure has not yet proved to be
feasible. As the F.S.N. global expenditure forecast is framed
within the Financial Bill (Budget) and this takes an excee-
dingly long technical time to be approved by the Parliament,
Regions have been allowed to adopt interim budgets to provide health services at the local U.S.L. level.
This procedure eventually implies that at the end of the
expenditure year, Regions 'bargain' with the central Government the approval of the sums already spent and which normally imply deficits. Until now, Regions' bargaining power has
always exceeded Government's resistance.
According to this mechanism, on the one hand expenditures
forecast at all spending levels present a high degree of uncertainty (e.g. F.S.N. Forecasts varied in 1982 between 4.5
and 5.8% of the G.N.P , · with obvious effects on decision-
making and allocative processes);on the other, effective
health expenditures have almost automatically been generated
that are largely out of central government's control.
Thus, the actual mechanism implies a substantial dichotomy
between the alleged 'structural' intent of public policy in
this sector, and the 'contingent' nature of its effective
action.
    Besides, no constraints have yet been devised to regula-
te systematically expenditures at the local level and U.S.L.
have proved to offer a weak resistance to doctor's and other
effective spending units' ability to consume resources.
It is widely recognized that the latter is, by and large, a
universal problem in health service systems, largely independent from specific institutional arrangements. In our case,
it may have been fostered by the absence (until very recently) of any charge collected at the point of service from
the individual patient-consumer, by the absence of control
and monitoring of doctor's productive performance, and by
the actual divorce between spending and taxing powers which
are located at different and distant levels of the allocative process.

3.2. Measures to Control Health Expenditures

    At present, different types of measures to enforce con-
trol and to ameliorate coordination at different levels of the
allocative process have been proposed and are under close
scrutiny on the part of the Government.
The following appear to be the most debated.
1) At the macro-economic level it has been suggested to

seperate the F.S.N. estimate and approval process from
the Financial Act (Budget) standard enacting procedure.
To shorten technical times, to decrease uncertainty and
to emphasize the structural content of the P.S.N, F.S.N.
financial amounts should be determined by the Parliament
through a specific 'health services financing act'.

2) Fund quotas to be allocated at the various U.S.L.,
should be determined at the central level to avoid shortcomings in health care provision otherwise due to existing
disparities in local availability of health care structures.

3) Any level of Regional health care expenditures in excess
of allocated quotas, should be met out of local taxation;
i.e. Regions should regain autonomous taxing power in this
respect, to avoid deficit spending and to establish a
closer connection between spending and taxing processes at
the local level.
Alternatively, an earmarked increase in personal income tax
rates has also been proposed, to reduce public deficit.

4) U.S.L. should monitor health care effective providers,
collecting systematic data on: doctors' remuneration; doctors' income allowances; doctors' rate of activity; pharmaceutical prescriptions per physician; prescribed lab. tests
per physician; rates of hospitalization; numbers of per
capita treatments; etc.

5) Regions should establish special Hospital Boards to assess
hospitalization needs and to avoid duplication in diagnostic analysis and other lab.tests.

6) Regions should finalize specific F.S.N. shares to professional training for ancillary personnel and for the managerial staff of the U.S.L.

7. To increase and extend the 15% charges on specialistic
treatments and lab-tests, as well as the proportional
charges on prescribed drugs, recently introduced to be
paid by consumers at the point of service.

8) To foster public initiatives aimed at increasing the
actual degree of awareness of the financial and economic
effects of medical action on the part of the Medical
Profession.

4.1. Innovation in Health Care Delivery Patterns

A full-length discussion of health services delivery patterns must be ruled out for brevity's sake and lack of syn-
thetic data.
Instead, a short description of the way in which the reformed
system could innovate the established pattern of PHC is presented, taking as an example the recently founded Consultori
Familiari (C.F. : Advice Bureau on Family Care).
It has already been mentioned how the present public

welfare system is expected to be reformed to be fully integrated with the S.S.N. and how U.S.L. should provide integrated social and health care.
A first step in this direction was the foundation of the
C.F. (1978/1979) on the part of Municipalities. These institutions, equipped with a multidisciplinary staff (e.g.
gynaecologist, psychologist, psychiatrist, pediatrician,
social workers, nurses) provide citizens with health care
and advice on various family issues. Although constrained
by financial and technical difficulties, the C.F. have proved
valuable in providing families with otherwise unavailable
assistance on : contraceptory techniques; psychoprofilaxis
for childbirth; maternal and children education; abortion and
breast-cancer prevention; etc.
At present, however, C.F. ranges of care appear unbalanced more in
favour of maternal care than global family care.

5.1. Problems in Management and Staff Organization

Within the newly founded system, management organisation at the local level is still far from being fully
satisfactory: although on the whole Regions' reactions to
Government's recommendations have been substantial, shortcomings and delays must be recorded in implementing new
functions and structures (e.g. about 90 out of 675 U.S.L.
still have to be established).
The Reform's highly innovative content has placed a great
burden on Regions' different resources and skills: in this
respect, Northern Regions have shown greater flexibility and
efficiency (e.g. in starting up regional plans; in devising
allocation formulas to share funds at the U.S.L. level; etc)
than Southern Regions. In some of the latter cases it has been
necessary to prolong the pre-reform Funds system because of
lack of adequate structures and staff.
Apart from purely technical or financial difficulties, the
actual level of managerial competence on the part of the
present staff may prove unsatisfactory in as much as it is
still rooted in the previous system's philosophy and roles. The
Regions' ability to implement government recommendations on
staff training, will prove crucial to ameliorate the present
situation.

5.2. Problems in Decision-making

Major general problems in the S.S.N. overall management
(apart from the already discussed financial ones) may be detected as far as the actual profile of the effective decision-
making process is concerned.

Synthetically, it may be observed that two problems are a major
cause of concern.
The first refers to the potential conflict between the 'indicative' nature of the planning procedure enacted at the
central government's level (and, consequently, the degree of
autonomy of choice and decision at the local level) and the
'structural' aims of public action in this field.
A good example may be provided considering the problem of
deciding how to allocate funds between the various U.S.L. over
a specific regional area.
According to one view, within the framework of indicative
planning, U.S.L. should deploy revenue according to autonomously detected local needs. But it may well be the case that,
for structural or contingent reasons, resource allocation
based on locally determined needs will be considered inefficient from an overall (social) point of view.
According to a different position (cfr. 3.2.) U.S.L. revenue
sharing should be completely determined at the central level,
and this may contrast with locally felt needs and with the
indicative nature of the planning procedure.

     A second major cause of concern is connected with the
task of conciliating effective decision-making units'
behaviour with the public interest. Although this seems to be
a fairly general problem (i.e. 'social' and 'individual'
benefits and costs may diverge in a purely private market
framework, as well), it appears crucial in public financed
settings.
Indeed, a relevant dichotomy present in the system is between
the effective units in charge of real spending decisions
(i.e. doctors, hospitals, etc.) and the public authorities in
charge of providing financial support to those allocative
decisions: as the fundamental motives and interest motivating
choices may diverge from collective aims, the actual quality
and quantity of health expenditures may well be publicly
financed but not necessarily publicly oriented.
Especially difficult in this respect seems to conciliate a
doctor's freedom of choice to define at the individual level
health needs and, consequently, to command resources with
financial limits and collective aims in the amount and composition of health care. Finally, it should be noted that
public control over health care providing agents (within its
effective limits) remains in several instances fragmented
beyond the health care framework: as an example, it may be
observed that U.S.L. power to modify health services personnel's productive performances is constrained by norms and rules
(e.g. collective bargaining agreements) which have, until
now, been set by public decision units outside the health
services system framework.

HEALTH SERVICES IN THE UNITED STATES: GROPING TOWARD A "SYSTEM"?

Anne R. Somers

Rutgers Medical School

New Jersey, U.S.A.

0.0. Introduction

There is no U.S. "health services system", in the formal sense
of the term, and very little formal coordination between the many
fragments, public or private, of this huge and vital industry, which
now accounts for nearly 10 percent of the Gross National Product
(GNP) and is estimated to cost about $275 billion in 1982.[1]/
In some important respects, there is even less coordination today -
under the antigovernment, pro-competition policies advanced by
the Reagan Administration - than there was five years ago. However,
this does not mean that the multitudinous programs, agencies,
instititions, and individual practitioners are totally lacking
in cooperative arrangements, coordination, or other "meaningful
relationships". On the contrary, there is widespread, albeit
inconsistent, evidence of increasing coordination, consolidation,
and even mergers between and among various institutions and
programs.

The purpose of this paper is to identify some of these
apparently contradictory trends and to try to determine whether
we are, de facto if not de jure, moving toward more systemization
or not. Does the current emphasis on competition represent a
permanent change in the long-run direction of U.S. health care
policy? Or merely a temporary pause for constructive realignment
of forces and adjustment to some profound changes in the underlying
supply and demand factors?

Limitations of space obviously preclude anything like a
comprehensive survey. I will simply attempt to identify

significant examples of the contradictory trends, starting with
goals and objectives, and generally following the requested outline.

0.1. Values, Goals and Objectives

The U.S. health care economy, like the general economy, is
struggling to reconcile two of the basic objectives set fourth
for the nation in the Preamble to the Constitution: to "promote
the general welfare", and to "secure the blessings of liberty to
ourselves and our posterity". Throughout most of our past history
the emphasis, in both the general and health economies, has been
on individual freedom and, in a frontier society with apparently
limitless resources, this worked reasonably well for most
Americans.

Almost from the beginning, however, there was recognition
that certain health problems could only be dealt with effectively
on a societal or "general welfare" basis. Thus, over the years,
there emerged communicable disease controls and other state and
local public health programs, governmental licensing of health
professions, provision of care for the mentally ill, workers'
compensation laws, etc. At this stage, our development was not
unlike that in Europe and, in 1912, about the same time that
Britain adopted its first national health insurance program, similar
demands began to emerge in this country.

Throughout the next half-century, interest in national health
insurance (NHI) waxed and waned, never commanding enough support
to overcome the opposition of the American Medical Association
(AMA) and other provider groups, but finally culminating in
Medicare, Medicaid (Titles XVIII and XIX of the Social Security
Act and our principal health care financing programs for the
elderly and the poor respectively) and a plethora of additional
health laws in the mid-1960s.
In 1966, the U.S. Congress adopted the Comprehensive Planning
Act (P.L. 89-749) which specified a commitment "to assure
comprehensive health services of high quality for every person".
For about a decade, this verbal commitment was taken seriously
and repeated efforts were made, by successive national
Administrations and most of the major interest groups, to formulate
some type of NHI program which would embody this commitment and
still prove viable in the highly fragmented American economic
and political environment.

The post-Medicare years were a great success in terms of health.
Almost all the relevant indices showed impressive improvement -
life expectancy at birth, life expectancy at 65, infant mortality,
progress with respect to control of most of the major diseases,
etc.2/But problems accumulated as well, most especially our
failure to control costs - the Achilles Heel of the entire

Great Society Johnson Administration program. In the 1980
Presidential election, the latter was, at least temporarily,
repudiated. In place of the goals of universal access to
needed health services and optimum quality, the de facto
goal today is purely and simply cost containment. Even
hospital spokesmen now proclaim the necessity of "shrinking the
system" and the method most frequently advocated is not
planning but elimination of the weak or unfit through competition.

The current debate is thus couched primarily in terms
of "competition" vs. "regulation", with the Reagan
Administration committed to the former, and supported - at least
for the first year of this Administration - not only by
physicians, hospitals and other providers, who chafed under the
increasing regulation needed to assure even our limited commitment
to universal access and high quality care, but also by taxpayers
who were called on to finance this commitment.

Significantly, however, as the moment of truth drew nearer
and the rhetoric of deregulation, "pro-competition", and
"consumer choice" began to be translated into specific laws and
regulations (sic!), disenfranchising increasing numbers of
potential patients and defunding programs to which providers, as well
as insurers, had become accustomed, second thoughts began to emerge.
Opposition to the new pro-competition proposals snowballed. At the
present writing, there is a legislative stalemate, reflecting the
unresolved conflict of values and objectives.

0.2. Basic Political Configuration

The reasons for the failure to translate an apparent
commitment into effective national policy are many and complex.
The following were especially important: (a) failure to control
the staggering rise in the costs of health care, especially after
1965 (see below) which convinced many economists, administrators,
and others that full realization of the goal of universality was
economically impossible; (b) the relative success of the fragmented
program, as reflected in improved health indices, thus blunting
the drive for full achievement; (c) the enormous strength of the
major provider interest groups, especially physicians and hospitals,
and their fear that a comprehensive government-controlled
system would lead to unacceptable controls over private activities
and earnings; and (d) the character of the American political
system.

Included under the latter heading were: (1) the separation
of executive and legislative powers in the American structure
of government, so different from the parliamentary system; (2) the
traditional fragmentation of federal and state legislatures; and
(3) the lack of ideological cohesion typical of our major political

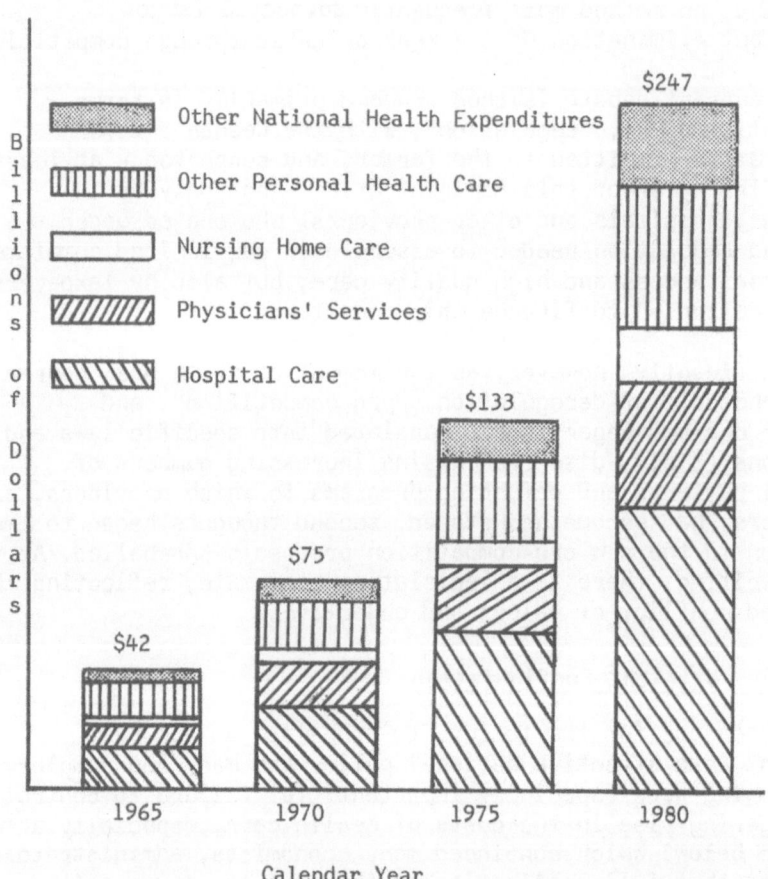

Figure 1   National Health Expenditures, By Type of Expenditure,
           Selected Years 1965-1980.   Source: Health Care Fin-
           ancing Review, September 1981, p. 5.

parties. In the absence of a clear sense of public crisis, and lacking party cohesion and effective government leadership, the various economic interest groups can exercise virtual veto power.

Most observers would probably add to this list of unique political characteristics our federal-state system, with its traditional concern for "states' rights". Certainly, this has been a major historical factor, slowing down the development of federal systemization in the health field as in many others. To this observer, at least, this now appears much less important than in the past. Witness the lukewarm state reaction to President Reagan's 1982 proposal for a "New Federalism". Most of our state governments now realize that "rights" carry with them "responsibilities" and an expensive price tag.

1.0. Health Services System Structure

The five following trends are especially significant.

a. Spectacular rise in overall costs. The "official" figure for 1980 is $247 billion, 9.4 percent of gross national product (GNP) (Fig. 1).3/ The latter may be compared with 3.5 percent in 1929 and 6.0 percent in 1965, the year that Medicare and Medicaid became law. Over the past 15 years, expenditures have grown at an average annual rate of 12.6 percent; between 1979 and 1980, 15.2 percent. The health care industry has clearly become a major factor in the overall U.S. economy, with impact and implications far exceeding individual health.

b. Increasing institutionalization. The changing distribution of national health care expenditures, by category of expenditure, 1965-1980, is shown on Figure 1. Within the dramatic overall rise, following are the major relative changes:

	1965	1980
Hospital care	33.3 %	40.3 %
Physicians services	20.2 %	18.9 %
Nursing home care	5.0 %	8.4 %
Other personal health care	27.1 %	20.7 %
Other national health expenditures	14.3 %	11.9 %
	———	———
TOTAL	100.0 %	100.2 %

The hospital and nursing home shares each rose substantially - the nursing home share by 68 percent - while all other categories fell proportionately. If construction costs are added to the institutional operating costs, their share of the 1980 national health care dollar would be over 51 percent. This does not mean that relatively fewer physicians services are being

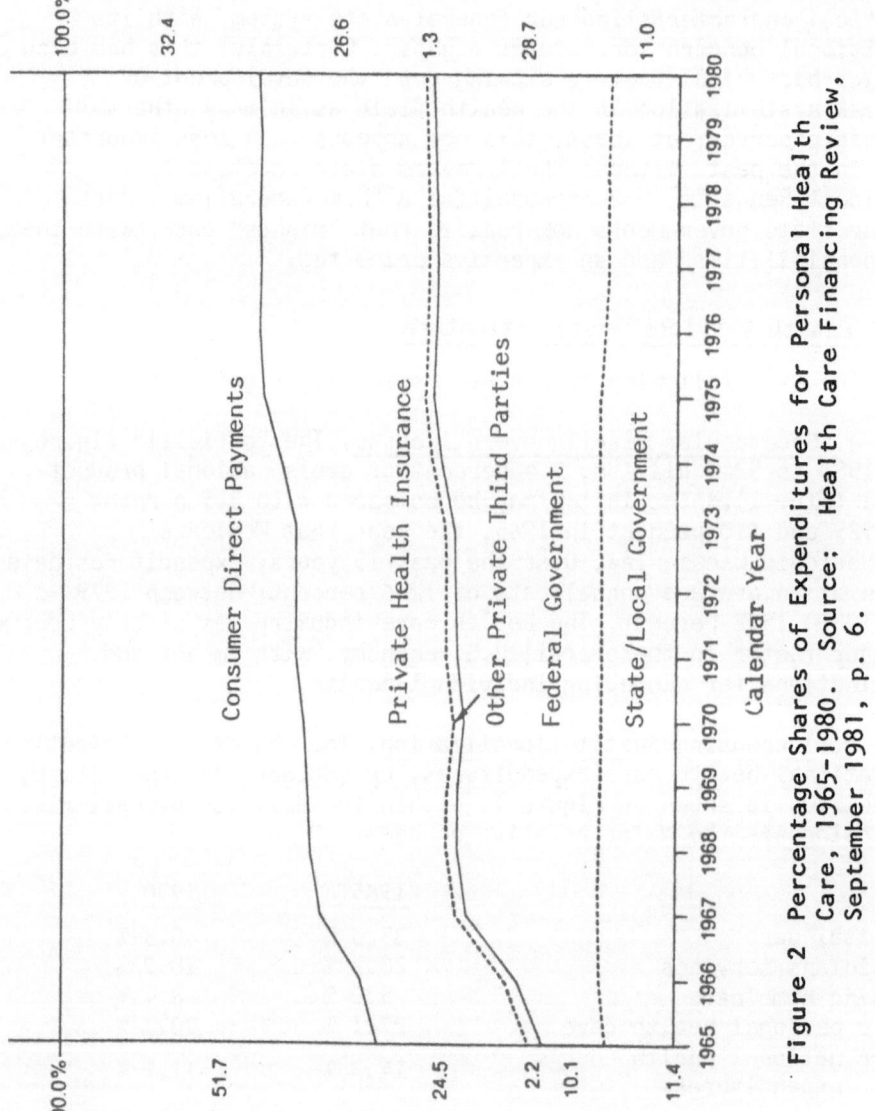

Figure 2 Percentage Shares of Expenditures for Personal Health
Care, 1965-1980. Source: Health Care Financing Review,
September 1981, p. 6.

rendered or fewer drugs prescribed. It does mean that relatively more of such services and goods are being provided in an
institutional setting. And an institutional setting is, or can
be, an important step toward systemization.

c. Increasing "third-party" payment. The term "third party" is commonly used in the U.S. to denote an institutional financial agent - either pulbic or private - who pays specified
providers for services rendered to specified patients in accordance with certain statutory or contractual agreements. Despite
the absence of NHI, the proportion of health care costs financed through such arrangements has increased dramatically since
the end of World War II. In 1950, about 35 percent of personal
health care expenditures 1) were paid through some form of
third-party; in 1980, 68 percent. At the same time, the proportion paid directly by patients out-of-pocket fell from 66 percent to 33 percent.

Private health insurance grew most rapidly during the
1950s and early 1960s; the public programs during the late
1960s following enactment of Medicare and Medicaid. Throughout
most of this period, the share paid by state and local government remained virtually the same as did the tiny share paid by
philanthropy and industry through direct services. Since the
mid-1970s the relative shares have remained quite stable, with
government responsible for the largest proportion, about 40
percent 2); private insurance about 27 percent; and consumer
direct payments about 1/3 (Fig.2).

d. Association between institutional care and third-
party payment. The extent of third-party coverage varies significantly according to category of care. As shown on Figure
3, 91 percent of all hospital care was paid through third-
parties in 1980; 55 percent through public programs. By contrast, only 63 percent of physicians fees were paid through
third parites; only 26 percent through government. The major
factors in this association appear to be the higher costs
and less predictability involved in institutional care. It
also suggests that third-party payment will continue to rise
as institutional care increases.

---

1) Total expenditures minus the costs of construction,
research, and certain public health programs. Personal
expenditures average about 88 percent of the total.
2) This does not include the substantial subsidy resulting
from tax deductions for medical expenses under federal income
tax law.

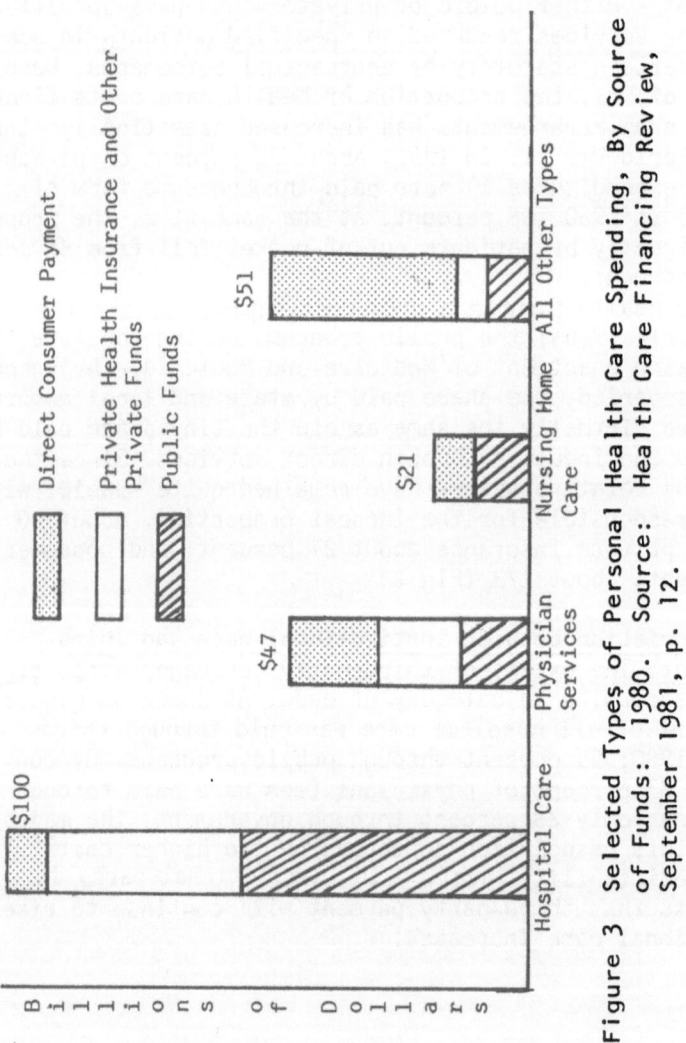

Figure 3   Selected Types of Personal Health Care Spending, By Source of Funds: 1980. Source: Health Care Financing Review, September 1981, p. 12.

e. <u>Price inflation as the major factor in the cost rise</u>.
Figure 4 shows the major factors in the cost rise for personal expenditures, 1965-1980. Approximately 58 percent of the
average annual increase was due to price inflation, as measured
by the fixed-weight price index for personal health care in the
Consumer Price Index. Another 9 percent is due to the growing
population. The remaining 34 percent is due to changes in the
mix of health services and goods purchased,in the frequency with
which these services and goods are used, and in the "intensity"
of care - for example, the number of tests performed on a
patient in the course of a particular surgical procedure. Whether
or not this "intensity" factor should be considered synonymous
with "quality" is a controversial issue among health care authorities. The importance of some form of price stabilization to the
future health of the health care system is apparent.

1.1. Coordination within the Health Sector

There appears to be a chicken and egg relationship between
the growth in technology, increasing institutionalization, high
costs, and third-party financing. Indeed there appears to be an
almost irresistable centripetal force in modern medical technology, with its high costs and high personnel requirements - a
force which is virtually independent of political ideology. Certainly all the trends noted above contribute both to the need
for, and the underlying elements of, a rational system approach
to the organization and financing of health care. Two examples
are illustrative. More will be provided in later sections.

a. Private Health insurance, Medicare, and Medicaid represent the typically American segmented approach to the health
needs of three target populations - the employed middle-class,
the elderly, and the poor. Despite this fragmentation, the
underlying centripetal forces continue to produce examples of
coordination: Medicare-Medicaid cooperation with respect to
certification of participating hospitals and nursing homes; coordination of benefits (COB) among private carriers of group
health insurance to avoid duplication of payment to patients
with multiple and overlapping policies; "Medigap" polices sold
by private carriers to fill the gaps in Medicare coverage resulting from deductibles and coinsurance; a new uniform claim form
for use by hospitals in billing Medicare, Medicaid and private
carriers.

b. The definition and establishment of physician fees and
institutional rates is clearly an area where professional self-
regulation is badly needed but probably illegal under existing
U.S. anti-trust law. Efforts in this respect, such as the "relative value schedules" developed by some state medical societies

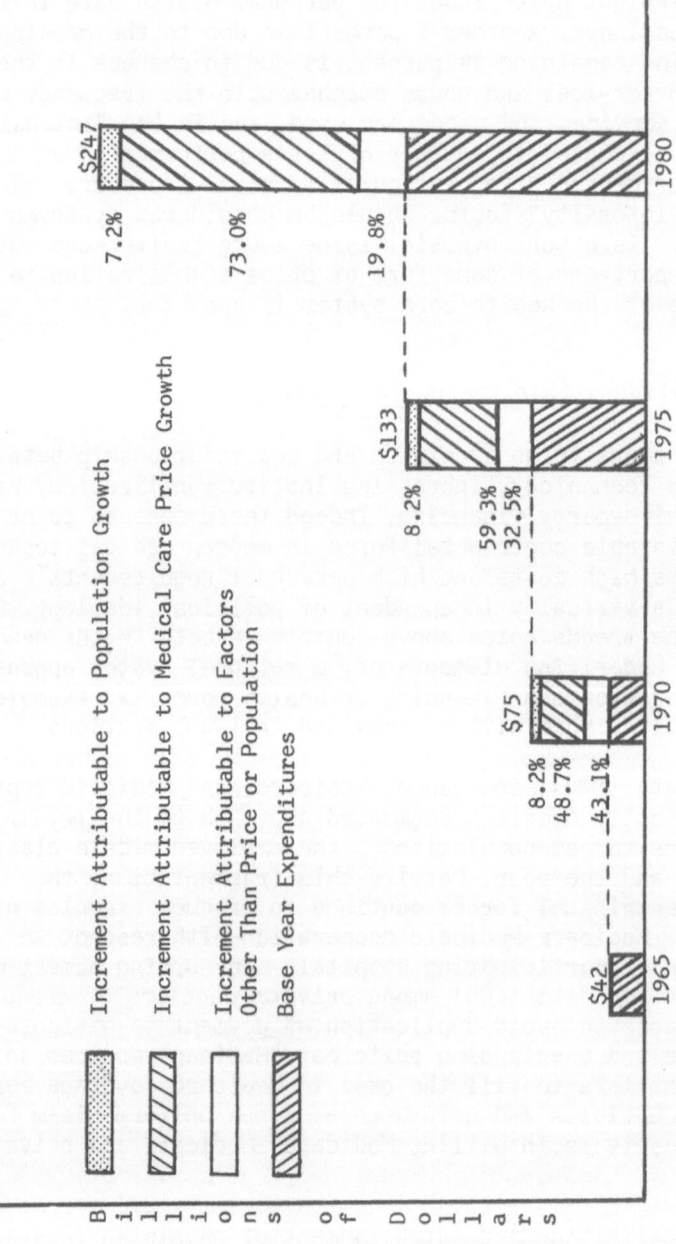

Figure 4 Factors in the increase of Personal Health Care Expenditures 1965-1980.
Source: Health Care Financing Review, September 1981, p. 8.

and specialty bodies, have been strongly attacked by the Federal Trade Commission and the Department of Justice, as inconsistent with anti-trust laws. On the other hand, efforts on the part of the AMA and the American Hospital Association (AHA) to restrain the rise in hospital costs through the Voluntary Effort (VE) Program launched in 1978, were generally lauded by government which suggests considerable ambivalence in national policy.

As a practical matter, the hospital merger movement, noted below, as well as the development of state rate regulation programs, are forcing coordination of rates as well as facilities among the affected hospitals. Anti-trust "principles" are likely to yield in this area.

1.2. Coordination with Other Sectors

Coordination between income maintenance and health programs was a key element during the Great Society years. Medicare was closely coordinated with the Social Security system; Medicaid with the welfare system. In the long and inconclusive debate over NHI, in the late 1960s and early 1970s, the question of continuing the tie between health insurance and Social Security became a major issue, with most private-sector organizations opposing such a tie.

Another conspicuous example of inter-sector cooperation has been in higher education. Most medical schools are organized within a university setting although there has been some recent movement toward free-standing schools. Leaders of the nursing profession are strongly in favor of academically-based educational programs, so much so that most hospital-based programs have disappeared.

One of the areas where cooperation is most urgently needed and most difficult to achieve is in the field of long-term care for the elderly and disabled. By definition such care usually involves patients who cannot be cured and who require long periods of care involving both medical and social services. To my knowledge, all countries, including those with highly centralized health care systems such as the British, continue to have problems of coordination in this area. These problems will not be solved simply by changing the source of payment. What is needed is a whole new approach to long-term care which recognizes that it inevitably involves a blend of medical and social skills and the need to develop both financial and administrative mechanisms to promote such a blending. The rapidly evolving fields of Geriatrics and Gerontology may contribute to this understanding not only with respect to care for the elderly but for the population in general.4/

2.0. Health Resources Production

2.1. Education and the Supply of Manpower

One of the oldest and most significant examples of non-governmental coordination is the complex network of testing, credentialling, and quality-monitoring bodies which, in the medical field, includes the National Board of Medical Examiners, American Board of Medical Specialties, the Liaison Committee for Medical Education, the Federation of State Medical Boards, and numerous additional boards and councils relating to the various medical specialties. Similar professional bodies dominate the fields of nursing, dentistry, podiatry, the rehabilitation therapies, and other allied health professions.

Over the past half century, these organizations have clearly contributed to an improvement in quality. At times, they insist on higher than public standards. Such private regulators also perform certain functions which, even under the tightest form of public coordination, would probably have to be delegated to them - for example, determination and policing of hospital privileges by "attending" (non-employed) physicians. Even if all hospital physicians were salaried by the institution, rather than the unique American "attending" system, some group of their peers would have to pass on eligibility and continue competence. There has always been some suspicion, however, that such controls could be exercised in a monopolistic manner - more for the benefit of the professions or institutions involved than the general public - and, in line with the current emphasis on competition, some branches of the national government are trying to break up this type of professional control.

2.2. Coordination of Facilities, Equipment and Supplies

This type of coordination has received a great deal of attention in recent years but the results are, at best, ambiguous. Efforts have proceeded along two entirely distinct lines: A) those aimed at geographic rationalization or "regionalization" of facilities and major equipment by means of public or voluntary planning, reinforced by public or quasi-public controls over capital expansion; and b) those aimed primarily at financial rationalization or consolidation of existing facilities through horizontal (mergers, development of hospital chains) or vertical integration (linkage of various types of health care institutions, rather than several institutions of the same type).

a. The health facilities (orginally limited to hospitals) planning movement emerged during the early 1960s with a variety of worthy, but frequently inconsistent, goals - enlargement and equalization of public access to health care, improved quality, cost containment, and, to some extent, the protection of privileged provider positions. Supported primarily by a series of federal laws - amendments to the Hill-Burton Hospital Construction Act of 1946, the Regional Medical Program (1965), Comprehensive Health Planning Act (1967), and Section 1122 of the Socail Securtity Amendments of 1972, requiring state "certificate of need" (CON) review of major capital expenditures by health facilities for purpose of Medicare and Medicaid reimbursement - the planning movement reached its apogée under the National Health Planning and Resources Development Act of 1974 (P.L. 93-641). Under the latter program, the nation was blanketed by a network of publicly-supported planning agencies at the local (Health System Agencies - HSAs), state (Statewide Health Coordinating Councils - SHCCs) and federal (National Council on Health Planning and Resources Development) levels. By 1980, some 210 HSAs were operative, covering virtually the entire country, and all states but one had CON programs, operating either under state law or under Section 1122.

The results are still controversial.5-7/ A few states could document both some restraints on capital expansion and improved distribution of facilities and major equipment . The concept of "regionalization" with its two essential components - (1) rational geographic distribution of facilities and programs, and (2) rational distribution of professional services encompassing the sepctrum of comprehensive care, primary, secondary, tertiary and long-term, with the affiliation agreements, referral patterns, and other connectives necessary to link the various levels and institutions into a coherent whole appeared to be gradually taking hold. 8/

However, the continued dramatic rise in overall health care costs, especially hospital costs, tended to discredit the entire planning movement. The fault lay less with what was done in the name of planning than with what was not done. The various health care financing programs were unwilling or unable to provide effective support through reimbursement controls. Medical education and the diffusion of medical specialists was totally at odds with the regionalization concept. The emergence of a new "school" of health economics - which attributed the intractable cost problem primarily to alleged "over-regulation" of the health care economy - provided a philosophical rationale for those who wished to pursue expansion regardless of long-run community or professional needs. Anti-trust laws were invoked both against planning agencies and third-parites seeking to support planning.9/ The Reagan Administration has opposed planning, both in principle and in practice, and has urged phasing out the entire federal effort. Faced with the real possibility that this might happen, the hospital industry is now widely divided on the subject. At the present writing, it is far from clear whether the federal law will be extended or not. In any case, many HSAs are already being dismantled and several states have repealed their CON laws.

    b. The consolidation movement had entirely different roots. Most of the early hospital chains evolved out of a common religious background, including several Catholic orders, the Mormons, Lutherans, and others. 10-13/ The Kaiser Foundation Health Plan has operated an effective chain of hospitals and ambulatory facilities for about 35 years. Several chains were able to demonstrate significant economies in terms of purchasing, insurance, personnel relations, shared management, and certain other areas. With the serious entry of investor-owned or for-profit hospitals into the health care picture, the movement toward chains mushroomed, especially in the South and West. By 1982, for example, Hospital Corporation of America - the largest of the for-profit chains - either owned or managed, on contract, 347 hospitals with over 50.000 beds, with total 1981 assets of about $3 billion, and 1981 gross revenue of $2.5 billion.13a/
    It is no accident that the investor-owned chains have been most successful in those parts of the country where planning and the certificate-of-need were least effective. During the past few years, however, the consolidation movement has taken on new urgency with the non-profits as well as the investor-owned institutions. With the intensified federal effort to cut back on hospital costs, the dismantling of the planning apparatus, the constant drumbeat of support for competition, and the threat of many hospital failures, even many successful non-profits are seriously considering merger or some other form of coordination. 14-15/ In 1982, for example, ten of the largest non-profit chains organized themselves into a new holding company, Associated Hospital Systems, representing some 150 hospitals and 30.000 acute care beds, either owned, leased or managed, plus an additional 33 institutions and 22.000 beds affiliated through some form of shared services. Total 1981 assets were estimated at $2.1 billion and gross patient revenue at $2.0 billion.16/ It is now estimated that the number of community hospitals will drop from about 6.000 to 5.000 by 1990 and that, of those remaining, 55-60 percent will belong to some form of multi-hospital system. 17/
    Vertical linkage is far from a new idea; it was advocated by the present author over a decade ago. 17a/ But, like mergers, it has suddenly acquired powerful new converts and many hospitals are busily acquiring, or developing affiliation agreements with, all sorts of health or health-related institutions - from nursing homes and home care agencies, to parking lots and trans-

port systems. Some, but not all, of this may be expected to benefit the public.

2.3 Coordination of Knowledge and Technology

Approximately $7 billion was spent on health-related research in 1980. 3/ Of this, about $2 billion was spent by drug companies and suppliers of health care goods and services. Of the remaining $5.4 billion, about 6 percent was performed by private non-profit organizations. Most of the other 94 percent - some $4.7 billion - was spent by the federal government, and most of this by the National Institutes of Health (NIH). Despite the categorical, disease-related, titles of most of the individual Institutes, the traditional NIH emphasis has been on basic biomedical, rather than clinical, research.

Over the past decade there has been increasing criticism of NIH for alleged failure to follow through adequately with the translation of basic research findings into clinical trials and applications. The multi-year, multi-billion War on Cancer was one outcome. In general, NIH resisted such pressures 18/ but, in fact, it has made concessions. The National Cancer Institute, for example, now has a Division of Cancer Treatment. The Heart, Lung and Blood Institute has an Office of Prevention, Education, and Control. The National Institute on Aging operates a Gerontology Research Center, headed by two directors - one Scientific, one Clinical. The latter Institute has also emphasized an interdisciplinary approach, including social scientists.

Not surprisingly, as one moves away from government research into industry, the competition becomes fierce. The stakes are high in both the drug and medical supply fields. Expenditures in 1980 for drugs, eyeglasses, and appliances were over $ 20 billion (not including hospital and other institutional purchases). 3/ Public efforts to monitor influence costs and quality in these areas have been ambivalent. As indicated above, there has been pressure to speed the release of research findings, sometimes even before the scientists were prepared to do so. On the other hand, every incident of inadequately tested products, such as the infamous thalidomide case, brings demands for more testing and more controls on premature marketing. Strengthening the authority of the Food and Drug Administration was one result.

More recently, highly sophisticated new technologies, such as open-heart surgery, renal dialysis, and the CT scanner, have been widely criticized as major contributors to the seemingly endless cost rise, with calls for better evaluation and control over their introduction and diffusion. 19-21/ Congress set up an Office of Technology Assessment to assess the biological, eco-

nomic, social,and political effects of technological changes in
a variety of areas, including health. 22/ Other high-level
efforts have been attempted but none have been very succes-
ful. The combination of public pressure for new "cures" combined
with the scientists' dedication to finding such cures and the
profits to be made from such cures, by all parties involved, has
thus far proved irresistable.

3.0. Economic Support of Health Services Systems

3.1. Coordination of Efforts to Control Total Costs

There is no serious effort to control total health care
costs. Such an effort was contemplated in the NHI proposal
supported by Senator Kennedy and organized labor during the
early and middle 1970s - one of the principal reasons for its
defeat. The many special interest groups which have proliferated
during the years of blank-check reimbursement have succesfully
resisted any effort to apply effective cost controls. Even the
Carter Administration's attempt to "cap" one segment of the industry - hospitals - in the face of 16-19 percent annual expenditure rises was successfully opposed by the hospital industry,
supported by the AMA. The industry's own Voluntary Effort (VE),
advanced as a substitute for governmental controls, did succeed
in moderating the rate of increase for 2-3 years. But as soon as
the threat of government action diminished, the VE effort
flagged and in 1981, hospital costs rose nearly 19 percent.
The basic problem, however, is not just lack of coordination
but the general absence of any form of effective cost control by
any third-party payor except the HMOs. The latter's relative
success may have actually contributed to confusing the issue. The
HMO "secret" is not necessarily capitation payment vs. fee-for-
service, but the facts that: (1) the less expensive primary
care services are emphasized as opposed to inpatient care, and
(2) all costs are effectively "capped" at least for a year at
a time.
In any case, the traditional "blank check" which most
third-parties have extended to most providers during the past 15
years is finally being withdrawn as Medicare, Medicaid, and even
some of the private carriers gradually introduce some limitations.
And, as already noted, in a few states, hospital rate regulation
is being developed. One way or another, cost controls will
inevitably emerge but we will probably have to live through
several more turbulant and difficult years of trial and error
before any generally satisfactory and acceptable methodology
is developed.

3.2. Coordination of Sources of Economic Support of Services and Research

Public Programs. As already indicated, there is very little coordination among the various governmental health programs at either federal, state, or federal-state levels. Developments of the past few years have been so contradictory that it is difficult to say whether there is more or less coordination than a decade ago. For example, the federal Public Health Service, which traditionally coordinated most of the categorical federal-state public health programs, such as maternal and child health, school health, Indian Health Service, etc., has been weakened as a result of the recent shift to "block grants" to the states. On the other hand, this shift makes possible, at least in theory, better coordination at the state level. Whether this will, in fact, happen - given the substantial cuts in federal funds that accompanied the change - remains to be seen. 23/

Similarly, bringing together Medicare and Medicaid - originally administered by two separate bureaus within the Department of Health, Education, and Welfare - into a single administrative unit, the Health Care Financing Administration (HCFA), could be construed as a positive step. This is also true of President Reagan's surprise proposal to federalize Medicaid if the states would assume responsibility for certain welfare programs. In fact, there has been little evidence of programmatic coordination between Medicare and Medicaid since the creation of HCFA and the two programs remain essentially as disjointed as ever.

Private Health Insurance. The private health insurance field is divided into three broad categories; the insurance companies, the Blue Cross and Shield plans, and the "independents". Although the non-profit Blue plans were first in the field, the insurance companies soon surpassed them in enrollment and have retained a majority position ever since. In 1979, the insurance companies accounted for about half of hospital, medical, and dental enrollment. The Blues had 39 percnet and the independents, a mixed group including HMOs and business-sponsored self-insured plans, accounted for the remaining 12 percent. 24/

For the most part, the different categories are highly competitive although de facto accommodations have evolved in some parts of the country. The insurance companies also compete with each other but this is rarely true of the Blue plans, most of whom operate in mutually exclusive geographic areas. A sort of "regulated competition" for employees of large employers has been encouraged for over two decades under so-called "dual choice" or "multiple choice" arrangements. The largest such plan is the Federal Employees Health Benefits Program (FEP)

which provides employees with choice of an insurance company
plan, a Blue plan, one or more HMOs if available where he or
she lives, and one or more other independent plans if
available. For the purpose of this program, the insurance
companies set up a special national consortium and the Blues
operate through the national Blue Cross and Blue Shield
Associations.

In the effort to strengthen the private sector and
discourage further government intervention, a few Blue
Cross plans and insurance companies have sponsored and helped
to finance fledgling HMOs, even though this involves competition with their own programs. A more common form of coordination involves the meshing of "major medical" insurance with
"basic" hospital and medical/surgical policies, regardless
of carrier. Coordination of benefit payments (COB) under group
policies between the two major types of carriers has been
noted.

Employers. About four-fifths of all health insurance
premiums are paid by employers, either from their own funds
or out of employee payroll deductions.* In 1980, this would
have generated around $200 billion. One might think this
would provide adeqaute leverage for significant policy impact.
In fact, most employers have been reluctant either to push
any special health care policy or even to coordinate action
for the sake of more efficient administration of existing
policies. The most conspicuous exception was Kaiser Industries
which was responsible for the innovative Kaiser Foundation
Helath Plan, the prototype HMO. 10/
Even the goad of rapidly rising health benefit costs has
not produced any really effective coordination. There has been
a considerable increase in articulated employer concern. In
1974, a new organization, the Washington Business Group in
Health, representing 200 of the largest employers in the
country, was established specifically to coordinate such
concerns and to head off any form of NHI that could mean
further government intervention in business. WBGH has advocated useful programs of various types, including more

---

* In 1975, group health insurance accounted for about 83
percent of total premiums and 96 percent of group insurance
was provided through employment and labor unions. A.K. Taylor,
"Employment Related Health Insurance", in Health-U.S. 1981,
Dept. of Health and Human Services, DHSS Publ. No. (PHS)
82-1232), p. 87

prevention, health education, and support for HMOs, and has
helped to spawn some 60 local or regional business coalitions
dedicated to the same purpose. 25/ A handful of these have
actually started to bargain with local providers. A few
employers have also actively promoted HMOs as more cost effective ways of delivering care.

At present, thus far the net effect of
employer action has been minimal.

Unions. Unlike employers, American unions have never
been inhibited in their advocacy of favored types of health
programs. As early as 1913, the International Ladies Garment
Workers Union (ILGWU) set up a Union Health Center in New
York City. A number of other labor organizations made similar
efforts. During the decade following World War II, the United
Mine Workers established an impressive network of hospitals
and clinics through the coal mining regions of Pennsylvania,
Wet Virginia, Kentucky, and Tennessee. The United Auto Workers
and others were strong advocates of prepaid group practice
long before the HMO concept became fashionable. The unions
played the key leadership role in the establishement and
diffusion of health benefits through collective bargaining and
then led the futile struggle for NHI. Other than physicians
and hospitals, no other group had such a lasting impact on the
size and structure of the U.S. health care economy.

At the present time, however, union influence has waned to
the lowest point since World War II. Although still able to
rally large numbers of workers to the defense of Medicare,
Social Security, and existing collectively bargained programs,
they are unable to provide effective leadership for any new
effort at significant systematic reform.

Other Sources. Of the four other sources of economic
support listed on the outline, one - private individuals and
households - still constitutes a major source of revenue for
the U.S. health care economy. As already noted, as late as
1980, comsumers were still paying nearly a third of our total
health care bill - $71 billion - directly out-of-pocket
(Fig. II) as well as a substantial portion of private health
insurance and even Medicare premiums. Almost by definition,
however, the out-of-pocket expenditures are uncoordinated.
Where consumers have purchased insurance that provides for
cost-sharing at the time of use through deductibles and co-
insurance, perhaps this could be construed as "coordination"
between individual and third-party payment. Cost-sharing
remains highly controversial, with the Reagan Administration
and most economists favoring it as a way of forcing consumers to be more cost conscious in their use of health ser-

vices while most providers, as well as most consumers,
oppose it, feeling that it discourages prevention and early
treatment, penalizes the poor, and raises administrative
costs.

The voluntary health agency is an important and unique
feature of the American health scene. The National Health
Council has a membership of 78 such organizations, including
the American Heart Association, the American Cancer Society,
American Diabetes Association, etc. These bodies have played
a significant role in public education with respect to their
particular disease and in lobbying for governmental support
of research and care in the various categorical areas. Several are effective fundraisers and provide some direct support
for care as well as research. Private philanthropic expenditures for personal health care in 1980 totalled about $1.4
billion. 3/ Together with industrial implant programs, they
accounted for 1.3 percent of total expenditures that year
(Fig. II). Despite their common membership in the National
Health Council, there is little evidence of any coordinated
efforts.

In the international field, the Department of Health
and Human Services has an Office of International Health,
headed by a Deputy Assistant Secretary. NIH has a unit -
The Fogarty International Center for Advanced Studies in the
Health Sciences - dedicated to this area. The Center for
Disease Control has an Office of International Services, etc.
The U.S. is a member of the World Health Organization, UNICEF,
and other bodies dealing with health affairs in other countries. It is not clear how much coordination there is among
these bodies but it is possible that there is more effective
coordination of the small resources we dedicate to international health than there is of the infinitely larger
resources within the U.S. itself. Such cooperation may break
down, however, when any substantial U.S. economic interest
is threatened, as it was in the infant formula case before
the WHO.

In line with the general growth of multi-national corporations, several U.S. hospital corporations now operate
overseas. The Hospital Corporation of America, for example,
owns or manages 15 hospitals with about 1,900 beds in the
Middle East, the Orient, Latin America, and Australia. 13a/

As to local community efforts, most major cities in the
U.S. have in recent years assumed responsibility for basic
public health services. Federal and state funds have been
channeled to local communities for this purpose through a
series of categorical grants, which, as noted, are now being
reorganized into block grants. Most of the larger cities
have also maintained public hospitals for the poor - many of
which are now in serious financial difficulty. For nearly two
decades, New York City - always in the forefront of public

health efforts - has made a valiant effort to coordinate
its 17 municipal hospitals. But, despite multi-billion dollar
budgets they remain on the verge of collapse. During the
1950s, Mayor LeGuardia took the initiative in setting up the
Health Insurance Plan of Greater New York (HIP), next to
Kaiser, the largest first-generation HMO.

An entirely different approach to community coordination
is represented by a new Robert Wood Johnson Foundation grant
program, "Community Programs for Affordable Health Care".
Cosponsored by the American Hospital Association and the Blue
Cross and Blue Shield Association, this $16 million program
offers grants for up to 10 communites to initiate projects
to contain health care costs.

4.0. Patterns of Health Services Delivery

4.1. Coordination of Health Services at All Levels

In theory most leaders of the U.S. health professions and
the health care industry, would probably agree on the desirability of an unbroken continuum of care from primary to tertiary or highly specialized, within designated regional boundaries
and with mutually agreed-on quality controls, well established
referral patterns, and adequate reimbursement arrangements to
permit patients to receive the most appropriate type of care
for their particular condition and to be transferred from one
level or modality to another as appropriate. In fact, the
situation is entirely different. The general failure of
"regionalization" with respect to facilities and equipment,
already noted, is reflected in the general absence of any
effective regional pattern of health services delivery.

In place of a rational division of labor between primary,
secondary, and tertiary programs, we have a situation where
most community hospitals and health facilites aspire to be
tertiary centerd and most physicians aspire to be specialists.
The reason is obvious: Specialism is weher status is, as well
as the bulk of reimbursement dollars.

The percentage of community hospitals with specific
services, as of 1980, was as follows:*26/

High technology services
CT scanners	22.6
Premature nursery	40.7
Hemodialysis, inpatient	24.1
Radiation therapy, megavoltage	16.3
Open-heart surgery facilities	10.7

---

* Based on reports from 91 percent of all AHA-registered
community hospitals

Low technology services
    Home care department   11.5
    Family planning   9.7
    Genetic counselling   7.2
    Alcoholism/drug abuse   14.5
    outpatient facilities

Perhaps there is justification for 11 percent of all community hospitals to maintain facilities for open-heart surgery and 41 percent to have premature nurseries. But surely there is no justification for the fact that less than ten percent have family planning services and less than 12 percent, home care. Basic to any serious effort at coordination or regionalization is broad distribution of the less expensive, more frequently used primary and preventive services coupled with more restricted distribution of the very expensive, highly specialized, tertiary services. Obviously our priorities are askew and unnecessarily high costs are just part of the prices we pay. Hopefully, the growing interest in "vertical integration" may help to correct this imbalance.

4.2 Primary Care : Function, Place, and Organization

The term "primary care" has recently achieved near-universal acceptance within health care leadership circles. Despite considerable inconsistency in usage, the term usually connotes general - rather than specialized - services, and continuity - rather than episodic - doctor-patient relations. It has also come to imply services that do not require sophisticated technology and hence should be less expensive. The popularity of the term derives from two quite separate but overlapping concerns: (a) popular yearning for reincarnation of the idealized family doctor of yesteryear; and (b) a conviction on the part of many health care authorities that a system in which primary care is emphasized will be both more cost-effective and health-effective than one in which patients have go start by self-diagnosis and then shop around for the specialist who appears to best fit their need.

The most important date in the development of primary care in the U.S. is 1969, when Family Practice emerged as a recognized specialty with a certifying board of its own, replacing the rapidly dwindling ranks of General Practice with a totally new breed of young and wll-trained physicians. Subsequently both Internal Medicine and Pediatrics began to stress primary or general care as well as subspecialty care. These three groups constitute the category of primary physicians (although on occasion OB/GYN is included) for pusposes of reimbursement, residency approval, etc. In the early 1970s, the precipitous decline in the ratio of primary

physicians to specialists ceased and there has since been a
near plateau. As of 1980, the ratio was 36 percent. 27/
     Interestingly, the proportion of medical school graduates
choosing residencies in one of the three primary care fields
is considerably higher - 57 percent in 1982. 28/ But the
realities of residency training and, especially, eventual
practice appear to cause many to leave primary care for one
of the specialties.
     The role of the federal government vis à vis the
strengthening of primary care has been highly inconsistent.
On the one hand, it has provided substantial financial
support for new academic programs and residencies and, for
several years, mandated that a specified percentage of
residencies be set aside for primary care - 50 percent in
1980. On the other hand, it has continued to discriminate
against primary care in Medicare and Medicaid reimbursement
formulas, especially in such areas as prevention and
counselling. This is also true of most private health insurance. HMOs are virtually the only type of PHI which
emphasizes primary care, although one or two individual
carriers have made some innovative efforts. This often-
neglected aspect of the HMO model is, in my view, one of
the principal reasons for its greater cost-effectiveness.
     Aside from HMOs, one of the few American institutions
with extensive experience in the sytematic coordination of
primary and speicalist care is the Hunterdon Medical Center,
a 200-bed community hospital in Flemington, N.J. The only
hospital in a large rural county, HMC opened in 1953 and
established from the outset a formal relationship between
the family practitioners who provided primary care in their
own offices and the specialists all of whom had offices in
the hospital and could only see patients on referral from a
primary physician. The latter turned over to the specialist
primary responsibility for their hospitalizes patients.
     For over twenty years, this arrangement worked well.
Although not an HMO - physicians continued to be paid on a
fee-for-service basis - their costs were in line with those of
Kaiser and significantly lower than other health insurance
in New Jersey. 29-30/ In 1979, however,the specialists
rebelled against the Hunterdon formula, and it is no longer
formally implemented. A practical and effective system was
sacrified to pressure for unrestrained competition.

5.0   Health Services System Management

5.1 Political Leadership and Managerial Skills

     Leadership at the top levels of the U.S. health care
economy is currently weak and inconsistent. The government
leadership is obsessed, almost exclusively, with cutting the

costs of public programs; the private sector leaders are resisting such cuts or at least those that affect their particular constituencies.

In the absence of any single health care system, it follows that we have no single Chief Executive Officer (CEO), Management Board, or other unitary administration. Formally, the broadest administrative control is in the hands of the Secretary of Health and Human Services (although de facto control lies beyond the Secretary in the Office of Management and Budget). With responsibility for Medicare, Medicaid, the National Institutes of Health, maternal and child health, and a broad range of other Public Health Service programs, the Secretary technically controls over 70 percent of all public health expenditures and about 30 percent of the nation's total health care expenditures. 3/ However, as a political appointee, with clearly limited tenure, and no authority over major federal programs such as the Veterans Administration, the Department of Defense, or state programs, such as Workers' Compensation, the Secretary's capacity for coordination is extremely limited. On several occasions during the past two decades, the suggestion has been made that we need some sort of overall Health Care Policy Board to report directly to the President, but all such suggestions were promptly shot down.

The same situation generally prevails at the state level. Authority over most health programs is usually divided between the Health Department and the Department of Human Services or Welfare. In New Jersey, for example, the latter department operates Medicaid and all state mental health programs. The Department of Higher Education controls professional education including the medical schools. Worker's Compensation is lodged in the Department of Labor and Industry. Community-based support services for the elderly are in a Department of Community Affairs.

Within the private sector, the tug-of-war between individual Blue Cross and Blue Shield plans and their central organization, the Blue Cross and Blue Shield Association, as to the latter's control over individual plans, has gone on inconclusively for years. The Health Insurance Association of America, trade association for the health insurance companies, and the Group Health Association of America, trade association for prepaid group practice plans (HMOs), exercise even less control over their members than does the Blue Cross Association. Whether all or any of these bodies might eventually emerge with some real authority will depend largely on the nature of the evolving health care system.

Among providers, each sector has its powerful trade association - the American Medical Association, American Nursing Association, American Hospital Association, etc. Aside from lobbying and public representation, these organizations have exercised control primarily with respect to education, credentialling, and quality controls (Sect. 2.1. above).

In January 1982, six major national organizations - the AHA, AMA, Blue Cross and Blue Shield Associations, the Business Roundtabel (parent of the Washington Business Group on Health), Health Insurance Association of America, and the American Federation of Labor-Congress of Industrial Organizations - issued an unprecedented joint statement reflecting their common concern for "the rate of increase in health care costs and the effects of public and private policies on the quality and access to health care." The statement supported establishment of local, state, and regional "coalitions", called for surveys of local resources and problems and particularly endorsed efforts at prevention, primary and long-term care. 31/ With no organization to implement its concern, however, the statement is likely to prove ephemeral.

Managerial skills are in great demand at all levels and command high salaries. Some of the new hospital chains have demostrated special competence in this respect. For example, the Hospital Corporation of American now holds management contracts with 146 U.S. non profit institutions and two over-seas. 13a/

Educational programs in health services administration and planning have proliferated during the past two decades. The Association of University Programs in Hospital Admini-stration reports a total of 66 U.S. graduate programs in 1982 plus 22 undergraduate. 32/ Nineteen were located in schools of business administration, 18 in schools of public health, 13 in general graduate education, 4 in medical schools, 4 in schools of allied health professions, and 8 elsewhere in the university. In the academic year 1979-80, there were 1.824 graduates who were promptly hired by hospitals, health insu-rance carriers government and other health and health-related agencies.

Health lawyers, accountats, management consultants, labor relations experts, and other professionals abound. But thus far there is little evidence that such a wealth of talent has made any significant contribution to the rationalization of the industry. On the contrary, the plethora of free-lance consultants probably constitutes a centrifugal force.

6.0   Conclusion

Even without passage of any of the much-advertised "pro-competition" Congressional bills, the current widespread retreat from planning and exaltation of competition, combined with the threat of a physician surplus and dollar shortage, have all contributed to weakening whatever elements of a

publicly-sanctioned regionalized health care system existed
a few years ago.

However, what may, in fact, be happening is not a rejection of rationalization as such, but a new approach,
featuring private enterprise, sophisticated financial management, corporate planning and mergers, rather than regional
planning and professional or public regulation.
The outcome is still far from clear. However, among the
numerous theoretical options two, at least, appear totally
unrealistic : (1) the possibility of reverting - perhaps via
exaggerated faith in the "unseen hand" of unfettered competition - to the highly fragmented health care economy of
pre-Medicare days, and (2) the possibility of moving rapidly
toward a single nationwide monolithic system under federal
control. By contrast, we are likely to move, by fits and starts
as we have in the past, through the grey area between the two
extremes, using a mixture of public and private enterprise, and
veering between the values of "the general welfare" and "individual freedom" and between emphasis on equity and quality, on
the one hand, and efficiency and economy, on the other. When
resources appear abundant, we tend to emphasize the former; when
resources are, or appear to be, scarce, the latter.

Since the coming decade is likely to witness a combination
that is surprising for the United States - an abundance of
physicians combined with a scarcity of health care dollars -
the results may also be surprising. But almost inevitably we
will continue groping toward systemization. The trend is simply
inherent in the centripetal force of modern medical technology.
32/ The trick will be to reconcile this with the changing
health care needs of an aging society dominated by chronic,
rather than acute, illness and disability.

References

1. W.K. Stevens, "High Medical Costs Under Attack as Drain on
   the Nation's Economy", New York Times, March 28, 1982.

2. Health U.S. 1981, U.S. Department of Health and Human Services, DHSS Publ. No. (PHS) 82-1232, Hyattsville, MD, December
   1981.

3. R.M. Gibson and D.R. Waldo, "National Health Expenditures,
   1980," Health Care Financing Review, vol. 3, September 1981,
   pp. 1-54.

4. A.R. Somers and D.R. Fabian, Eds., The Geriatric Imperative:
   An Introduction to Gerontology and Clinical Geriatrics, New
   York City : Appleton-Century-Crofts, 1981.

5. C.C. Havighurst, Ed., Regulating Health Facilities Con-
   struction, Proceedings of a Conference on Health Planning,
   Certificates of Need, and Market Entry, Washington, DC:
   American Enterprise Institute, 1974.

6. D.S. Salkever and T.W. Bice, "Certificate of Need Legisla-
   tion and Hospital Costs", in M.Zubkoff, I.E. Raskin, and
   R.S. Hanft, Hospital Cost Containment: Selected Notes for
   Future Policy, New York : Prodist for Milbank Memorial
   Fund, 1978, pp. 429-460.

7. National Academy of Sciences - Institute of Medicine,
   Committee on Health Planning Goals and Standards, R. Fein,
   Ch., Health Planning in the U.S. : Selected Policy Issues,
   Vol. I and II, 1981.

8. E.Ginzberg , Ed., Regionalization and Health Policy, U.S.
   Department of Health, Education, and Welfare, DHEW Publ.
   No. (HRA) 77-623, 1977.

9. W. Greenberg, Ed., Competition in the Health Care Sector:
   Past, Present and Future, Proceedings of a Conference Spon-
   sored by the Bureau of Economics, Federal Trade Commission,
   Washington, DC, 1978.

10 A.R. Somers, Ed., The Kaiser-Permanente Medical Care Program:
   A Symposium, Oakland, CA: Kaiser Foundation Health Plan,
   Inc., 1971 (rev. 1979).

11 Multiple Hospital Units under Single Management. Report of
   the 1965 National Forum on Hospital and Health Affairs,
   Durham, NC: Graduate Program in Hospital Administration,1965.

12. B.J. Jaeger, Ed., A Decade of Implementation: The Multiple
    Hospital management Concept Revisited . Report of the 1975
    National Forum on Hospital and Health Affairs, Durham, NC:
    Department of Health Administration, Duke University, 1975.

13. J.P. Cooney and T.L. Alexander, Multi-hospital Systems: An
    Evaluation. Chicago: Hospital Research and Educational Trust
    and Northwestern University, 1975.

13a Hospital Corporation of America, Department of Corporate
    Communications, Nashville, TN, telephone communication, May
    7, 1982.

14. B.J. Jaeger, Ed., Vertically Linked Health Organizations.
    Report of the 1978 National Forum on Hospital and Health
    Affairs, Durham, NC: Department of Health Administration,
    Duke University, 1978.

15. H.S. Zuckerman and L.E. Weeks, <u>Multi-Institutional Hospital Systems</u>. Chicago: Hospital Research and Educational Trust, 1979.

16. Lorna Lafko, Assistant to the President, Sisters of Mercy Hospitals, Inc., Farmington Hills, MI, telephone communication, May 8, 1982.

17. <u>Hospitals</u>, Journal of the American Hospital Association, vol. 56, April 1, 1982, p. 23

17a A.R. Somers, "Rationalization of Community Health Services and the Role of the Hospital," in A.R. Somers, <u>Health Care in Transition: Directions for the Future</u>, Chicago: Hospital Research and Educational Trust, 1971, Chap. 7.

18. President's Biomedical Research Panel, <u>Report Submitted to the President and the Congress of the U.S.</u>, U.S. Department of Health, Education, and Welfare, DHEW Publ, No. (OS) 76-500, Washington, DC, 1976.

19. L.B. Russell, <u>Technology in Hospitals: Medical Advances and Their Diffusion</u>, Washington, DC: Brookings, 1979.

20. S.H. Altman and R. Blendon, Eds., <u>Medical Technology: The Culprit Behind Health Care Costs?</u> Proceedings of the 1977 Sun Valley Forum in National Health, U.S. Department of Health, Education, and Welfare, DHEW Publ. No. (PHS) 79-3216, Washington, DC,: 1979.

21. A.D. Spiegel, D. Rubin, and S. Frost, <u>Medical Technology, Health Care and the Consumer</u>, New York: Human Sciences Press, 1981

22. Congress of U.S., Office of Technology Assessment, <u>Assessing the Efficacy and Safety of Medical Technologies</u>, Washington, DC, 1978.

23. G.S. Omenn and R.P. Nathan, "What's Behind those Block Grants in Health?" <u>New England Journal of Medicine</u>, vol. 306, April 29, 1982, pp. 1057-60.

24. M.S. Carroll and R.H. Arnett, "Private Health Insurance Plans in 1978 and 1979: A Review of Coverage, Enrollment, and Financial Experience," <u>Health Care Financing Review</u>, vol. 3, September 1981, pp. 55-87.

25. J.K. Iglehart, "Health Care and American Business," <u>New England Journal of Medicine</u>, vol. 306, January 14, 1982, pp. 120-125.

26. Hospital Statistics, 1981 Edition, Chicago: American Hospital Association, 1981, Table 12A, pp. 190-197.

27. C.M. Bidese and D.G. Danais , Physician Characteristics and Distribution in the U.S. 1981, Chicago: AMA, Division of Survey and Data Resources, 1982 (in press).

28. August Swanson, Director of Academic Affairs, Association of American Medical Colleges, Washington, DC, telephone communication, May 7, 1982.

29. H.B. Curry, et.al., Twenty Years of Community Medicine: A Hunterdon Medical Center Symposium, Frenchtown, NJ: Columbia, Publ. Co., 1974

30. L.B. Wescott, "Hunterdon: The Rise and Fall of a Medical Camelot," New England Journal of Medicine, vol. 300, April 26, 1979, pp. 952-56

31. J.T. Dunlop, Coordinator, Lamont University Professor, Littauer Center, Harvard University, Cambridge, MA, Press Release, January 1982.

32. Director of Research, Association of University Programs in Hospital Administration, Washington, DC, telephone communication, May 5, 1982.

33. A.R. Somers, "Health Care Technology and the Political System: The Sorcerer's Apprentice Revisited," in A.R. Somers and H.M. Somers, Health and Health Care: Policies in Perspective, Germantown, MD: Aspen Systems Corp., 1977, pp. 479-490.

THE HEALTH SERVICES SYSTEMS IN BELGIUM

Jan E. Blanpain

Centrum voor Ziekenhuiswetenschap

Leuven University, Belgium

Introduction

The current Belgian health care system was shaped in the post World War II period. The government inspired by the developments in Great Britain saw it as its role to organize solidarity and to seek income redistribution for security against unemployment, against loss of income through prolonged illness or expenditure for medical care, in case of retirement, etc. A rather ambitious social security program was thus launched, eventually making the whole population eligible for sickness insurance, unemployment benefits, family allowances, retirement benefits, invalidity bene- fits. Moreover salaries, wages and income replacements became subject to cost of living adjustment.

Important to note is that the existing providers and the existing private sickness funds became after World War II the channels and carriers of the program of comprehensive social se- curity coverage.

The at that time existing health care system and existing private sickness funds gradually became involved in a loosely structured but heavily regulated set of arrangements to provide care at "zero-cost-at-the-time-of-service", financed through pay- roll contributions and through state subsidies.

Free choice of patients, clinical autonomy of physicians and unrestricted access to higher education are highly cherished values. Infringement inspired by spatial, administrative, tech- nological, financial or other rationalities are fiercely resisted to the extent that leftist parties prefer to forego health man-

power planning to safeguard free access to education. Another
paradox related to these values is the insistence of the medical
profession, traditionally wary of government intrusion, of
stringent government planning of resource production and deployment.

Solidarity among citizens is constantly being persued by
the leading political parties although there is now serious controversy over the mechanisms which seem to favor the well-off
at the disadvantage of the underprivileged. In a well documented
and detailed study, H. Deleeck recently proved that the Belgian
social security system is biased towards the middle class (1).
His study echoes the findings of the Black report in the United
Kingdom (2).

Finally pluralism in the provision and financing of health
care and a delicate balance of public and voluntary endeavours
are constantly to be taken into account.

Belgium as a parliamentary democracy is currently undergoing a far reaching process of federalization. Three sub-national
entities emerge: Flanders, Wallony and Brussels. The former two
already have separate legislative and executive bodies and
separate budgets with autonomy in such matters as housing, industrial development, culture, health and welfare, environment. The
status of Brussels is still being negotiated. It looks as if
such matters as education and social security financing, including
health care financing could shift in the near future from the
national to the subnational level.

Health Services System Structure

The health care delivery system is composed of independent
private practitioners in primary care and a mix of public and
voluntary institutions at secundary and tertiary care levels.
For-profit private institutions are a negligible minority.

General practitioners and to a large extent specialists provide care in surgeries on a solo-praxis basis. Group practices do
exist but are a minority sofar. Specialists also function in free-
standing and hospital-based outpatient departments. Patients have
direct access to specialists. Home nursing services in liaison
with G.P.'s are well developed. There is already an oversupply of
manpower in aggregate terms.

Acute hospital care, is more or less evenly divided between
a voluntary religious sector and a public local authority sector.
Many cities have a hospital of each sector. This tends to create
problems of coordination although examples of public-voluntary
multi hospitals systems do exist.

Extended hospital care is, contrary to many other countries, predominantly provided by voluntary organizations. Psychiatric inpatient care is in the process of being reorganized and sharply reduced in numbers of beds per institution and in terms of total beds. Somatic extended care, in particular care for the elderly has lagged behind in its development regardless of the demographic evolution.

The health insurance system is compulsory but is carried through non-for-profit private sickness funds. Payroll deductions, employer contributions and state subsidies finance the health insurance. Reimbursement of negotiated fees or per diems, plus cost sharing by patients, finance the providers. Both private and public providers receive the same financing with exception for the public institutions who receive a higher degree of state subsidy for capital investment or coverage of operating deficits.

By and large the money flows of the social security system have been indirectly used to guide, without an overall strategy, the deployment and expansion of the health care delivery system through granting or withholding reimbursement for given types of care.

Until 1970 the coordination of the different components of the health service system took place within a growth model. State subsidies were used to promote the building of hospitals and of health screening centers.

Around the mid-seventies planning legislation emerged which since subjects capital investments, including the instalment of costly technology, to a certificate-of-need regulation within a set of decreed norms. An Advisory Body at the national level determines the norms while committees at the subnational level examine the individual capital investment requests. The former advises on norms where the latter advise on specific requests. The whole coordination effort sofar has however been mainly centered on hospital facilities and equipment with neglect of other resources. At the time of this writing the entire advisory machinery for health facilities planning was abolished by the government which operates with special powers thus bypassing parliament. A new more integrated body has been announced. Meanwhile the government has stopped all health facility construction. It also acquired the power to close facilities which are not in accordance with the decreed norms per population.

Health Resources Production

Health manpower production in Belgium is virtually uncontrolled. Educational opportunities in every direction at extreme low cost are practically guaranteed by the Consitution and by a

tradition which seeks to overcome social gradients in the student
body. A long history of occupation by foreign powers with discrimination of certain groups as a consequence has resulted in a great
sensitivity for measures which might discriminate in access to
education. This and unlimited enrollment beyond educational capacities are producing an oversupply of doctors, nurses, pharmacists, physiotherapists, laboratory technicians, etc. The production
of medical specialists has recently been curbed by a limitation
of the number of training positions in hospitals. The deployment
of pharmacists is subject to a permit and a vacancy within a
fixed number of positions in the country. Organized medicine is
lobbying intensively for a "numerus fixus" in education and for a
limitation of positions accessible in the health system. Educational
institutions however are reluctant to abandon a financing system
based on student-intake. Political parties in favor of open admission are being accused by the medical profession to pursue a
hidden policy of income erosion.

Physical resources such as hospitals, technical equipment
and drugs are subject to production approval. With exception of
drugs where scientific investigation and valid criteria of efficacy,
safety and efficiency are used in the approval process, the physical
facilities are approved on the basis of questionable criteria and
a partisan approach by those in power.

The production of knowledge in response to the needs of the
health care system is characterized by poor research management
and by a lack of follow-up. Major sums were committed to research
in the field of primary care and in the application in the seventies
of computers in health care (3). The former led to a sterile ideological debate of partisan sociologists and the latter petered out
in funding operational applications of existing know-how. Health
services research has been seriously neglected because of lack of
support. Several Ministers of Health either ignored research or
simply denied resources for health services research.

Economic Support of Health Services Systems

Faced with the worst economic crisis since 1930, the Belgian
government is bent on cost containment of health care. It is such
a priority in government policy that a revision of the total
security system is under way. A serious shift in the public/private
mix of health care financing such as higher premiums for insurance,
less coverage and more cost sharing can be expected. The measures
to control total costs of health care are however uncoordinated:

- Cost sharing for primary care recently introduced will probably
  counteract a desire to shift inpatient care toward primary care.

- A reduction of hospital bed supply is being implemented through
  penalizing patient days beyond given limits. A 3% reduction is

thus being pursued for 1982.

- Replacement of acute hospital beds by extended care facilities is considered but implementation lags because of naive concepts about extended care.

- Cost sharing for drugs depending on the degree of drug efficacy has been introduced.

- Reduction in fees for laboratory examinations and X-ray examinations have been drastic bringing hospitals dependent on such income to the brink of bankruptcy.

- Introduction of hospital budgets below current spending levels is being announced.

- Utilization review of physician performance and hospital performance to eliminate fraudulent abuse has started. However the first major effort using profiles of physician performance proved to be totally unreliable leading several physicians accused of impossible abuse - like dentists doing deliveries - to sue the government.

Patterns of Health Services Delivery

The Belgian health care system is characterized by a separation (administratively, physically, financially and staffing) of preventive and restorative services.

Within the preventive domain there are different agencies without any coordination: Mother and Child Health; School Health; TB-control; Industrial Health; Sport Health; Cancer Detection; Health Education; Mental Health; Family Guidance; etc.

The Ministry of Health sofar has proven incapable to integrate the myriad of overlapping preventive services. A bill to streamline preventive health care and to focus on preventive practices of proven effectiveness has been inactive for almost ten years. Within the restorative sphere, primary care is provided by independent general practitioners, dentists, pharmacists, competing home nursing services, physio-therapists and directly accessible specialists.

There is little systematic support of primary care organized by higher functional levels of care. Direct access by patients of secundary and tertiary levels results in common problems being addressed at inappropriate levels.

Health Services Systems Management

The Belgian health care system fundamentally lacks a strategy. Lipservice is being paid to the need for a national health policy, the desirability of a structured delivery system with clearly defined roles of each tier and well defined referral procedures.

A major obstacle to a rationalized and systematic approach to health care is the absence of a health information system and the neglect of epidemiology as a field and as a practice. Information on health status, health resources, health care utilization, health care expenditure, health care costs and health care impact is either non-existent or unreliable or piece-meal.

The lack of information is compounded by the absence of consensus among the major parties involved regarding basic structural and functional options towards which the existing system should be changed.

References

1. Deleeck, H. et alii, De Sociale Zekerheid, Tussen Droom en Daad, Van Loghum Slaterus, Deventer 1980.
2. Townsend, P., Davidson, N., Inequalities in Health, Penguin Books, Harmonsworth, 1982.
3. Blanpain, J., Delesie, L., Community Health Investment, Health Services Research in Belgium, France, Federal German Republic and the Netherlands, Oxford University Press, 1976.

THE HEALTH SERVICES SYSTEM IN TURKEY

M. Rahmi Dirican

Department of Public Health
Medical School
University of Bursa, Turkey

Introduction

1.1. Geography and Climate

The Republic of Turkey covers 756.000 km^2 of the Anatolian peninsula in Asia, and 24.000 km^2 of the Thracian peninsula in Europe. It is both geographically and historically, a link between East and West. This influences the entire fabric of life. Turkey is a mountainous country, the average elevation in Central Anatolia is around 1.000 meter and in Eastern Anatolia is around 1.950 meter. Being situated in the temperate zone, Turkey has various climatic types in different parts of the country. The average annual temperature varies between 18-20°C on the South Coast, falls to 14-15°C on the West Coast and in the interior areas fluctuates between 4-18°C. The East Anatolian and the interior parts of Turkey are subject to cold winters and average temperature over these areas are between 0 and -10°C in winter. Turkey is under influence of a continental type climate characterized by rainy weather throughout the year and also of a sub-tropical climate distinguished by dry summers. It almost never snows in the South and West Coasts, on the contrary East and North-East parts of Turkey are under heavy snow at least four months a year.

1.2 Population

According to the 1980 census (1), the population of Turkey is 44.737.000. Fifty six percent of the population lives in the rural areas. There are 36 cities over 100.000 population and 34.678 communities under 2.000 population. The average population of villages is 694. The population density varies between 16

(Province of Hakkari) and 829 (Province of Istanbul) per square
kilometer.

The percent distribution of population in three broad age
groups is as follows:

Age groups (in years)	Percent of population
0-14	39.9
15-64	55.3
65$^+$	4.8

Source: State Institute of Statistics
Population census of Turkey 1975
1% sample results, publication
no. 771, 1976, p. 1

The annual rate of population increase in 1940-1944 period
was 10.6 per thousand. It reached 28.5 per thousand in 1955-1959
and dropped to 20.7 per thousand in 1975-1979 period.

1.3 National Income

Turkey's gross national product in purchaser's prices in 1980
is estimated at 4.433 billion Turkish liras (almost 44 billion US
dollars) and per capita national income at current factor cost was
88.983 Turkish liras (2), (or 890 US dollars). Industry constitutes
23.9 percent of the gross national product. The other noteworthy
sectors according to their contribution to the gross national product were agriculture (21.1 percent), wholesale and retail trade
(14.7 percent), and transportation, storage and communication (9.3
percent).

1.4 Health Situation

Precise morbidity and mortality data are almost always a by-
product of a highly developed medical system. It follows that
health information in a developing country with inadequate health
services can be expected to be scanty and approximate. On the
other hand, physicians and other health personnel attach little
importance to reporting vital events. In Turkey deaths are reported more reliable than births, except in villages, where both
are grossly under-reported. Under these conditions for an accurate
definition of the health situation special surveys are needed.
Only a few such studies have been done in Turkey. The following
vital statistics are obtained from different sources (3,4):

Birth rate (per 1.000 population)   30
Death rate (per 1.000 population)   10
Natural increase (%)   2
Infant Mortality rate (per 1.000
                        live births)   ± 140

1.5 Goals and Objectives of the Health Services System

In Turkey, to improve the health conditions of the country,
to combat all diseases and other harmful factors detrimental to
the health of the population, to ensure healthful living for
future generations and to provide medical and social assistance
for the people are among the responsibilities of the State (5).
The Ministry of Health and Social Assistance (MHSA) is the organ
through which these responsibilities are rendered. The MHSA is
the highest organ and supervising authority in respect to the
health services in the country with the exception of the health
services operated by the Ministry of Defense. Services left in
the charge of the local and municipal authorities come under
the supervision and control of MHSA (5).

Health services in Turkey passed through several stages of
evolution since the formation of the republic in 1923. From that
time until the 1950's state-provided health services and preventive
medicine had the priority. Because the priorities and policies
were set correctly, Turkey was able to overcome many of the prevalent communicable diseases, although there were great constraints
in financial and manpower resources.

From 1950 onwards, along with increased rates of urbanization
and industrialization, emphasis shifted to medical rather than
health care. Many hospitals were built and private practice began
to flourish. Rural population and the urban poor living in shanty
towns were neglected. Health indices began to deteriorate. Demand
for medical care increased and environmental conditions became
worse. The organizational set-up founded in the early years of the
republic was not able to cope with the increasing demands and needs.

To overcome these deficiencies a major policy decision was
taken in 1961 and the nationalization of health services was initiated from the Eastern provinces. Under the regulations of the
Turkish National Health Services, physicians and other health personnel who work in the district receive a fixed salary and do not
have the right of private practice. The basic units of nationalized
health services are multi-purpose health units. A unit serving
less than 10.000 population is staffed with one physician, two
public health nurses, three to five auxiliary nurse-midwives (ANMs)
and one medical secretary. In the district, each such team serves

3.000 to 9.000 persons, and each ANM serves approximately 2.000
persons. ANMs in the rural areas reside in the health station in
the central village of their area. In units serving more than
10.000 population, the staff is increased accordingly. The physician, with his paramedical workers, practices comprehensive
medicine. They are responsible for ambulatory and home care, for
all preventive work and for supporting community development
activities. Their superiors are the provincial health directors
who are responsible to the Governor for directing and supervising
health services in the Province. The activities of the medical
officers in the Health Units are supported by hospitals, dispensaries and public health specialists.

Although the nationalization of health services has expanded
over the years and covers 45 out of 67 provinces now, the system
does not work properly as it was planned. Deep-seated bureaucracy
of MHSA, economical difficulties, powerful pressure groups and
short-sighted politicians have all played their roles in hampering
this progressive re-organization of health services. To run efficiently, the nationalized health services should be decentralized;
top level administrators must have management skills; teaching
and training of physicians and paramedical personnel must be relevant to the needs of the country; larger parts of the government
budget have to be alloted for the MSHA; health personnel should
be evenly distributed all over the country and vertical health programmes must be integrated.

At present, expansion of the nationalization of health services has practically stopped, and health policies are mainly
geared towards enlarging the private sector. The budget of MHSA
has reached a minimum low level. Community care and state medicine
have lost their priorities. The general health status of the public
has become worse not only because of a decline in the health
system as a whole, but also because of decrease in real income
and purchasing power, deterioration of housing conditions and nu-
tritional status of the people. In order to increase the effectiveness of health services some steps have been taken as well.
Salaries of health personnel have been increased a little and integration of some vertical programmes such as tuberculosis control
and family planning being attempted. A compulsory service of
two years for medical school graduates and for specialists has
recently been accepted by the Government.

Recent political disturbances had unfortunate impacts upon the
country. Increased terrorism and political conflict ended up in a
military take-over. The military intends to consign to a civilian
democratic government as soon as possible. A new Constitution is
under discussion in the Consultative Assembly. The election will
certainly be held before the end of 1983.

II. HEALTH SERVICES SYSTEM STRUCTURE

2.1 Health Care Providers

There are five main providers of health care in Turkey. The first one is the MHSA which is responsible for improving the health conditions of the country, with implementation delegated to municipal and local health authorities. Its activities divided in the traditional categories of public health and medical care services. The public health services are strictly the responsibility of the MHSA. The administrative organization of the Ministry of Health is divided into two parts - central and provincial. The Ministry of Health is ultimately responsible for all health services. The Undersecretary is the chief executive officer and handles routine administration. Assistants to the undersecretary are in charge of various directorates such as tuberculosis control, family planning and personnel.

Each province has a Director of Health appointed by the Ministry of Health. He is the highest health official in the province but is officially under the administrative control of the governor. Each province is divided into kazas or districts with a kaymakan as the highest administrative officer. In every kaza there is a medical officer of health who acts as health advisor to the kaymakam. In cities medical officers work directly under the Provincial Director of Health but in the kazas they are under the administrative control of the kaymakam. The Ministry of Health appoints municipality doctors but they are paid by the municipal authorities.

Drawbacks of this central-authoritarian type of administrative structure of the MHSA has been repeatedly discussed but no attempt has been made so far to change it.

Since all the other providers of health care are dealing with only medical care, their relative responsibilities for medical care could be based on their provision of hospital beds (Table 1). Using the criterion, 58.6 percent of the medical care services are supported by the MHSA and 2.7 percent by the municipalities and local authorities. The limited contribution at the municipal and local level is due to poor organization and lack of funds. The MHSA has the authority to supervise the administrative arrangements and health standards of all health institutions except those operated by the Ministry of Defense.

The relative responsibilities of main providers of health care could also be based on the number of physicians working in their organizations (Table 2). As it is seen nearly one quarter of physicians are working in the MHSA.

Table 1 - Bed Capacity of Health Institutions by Main Health
           Care Providers (1980)

Health Care Providers	Bed Capacity	
	No. of beds	%
MHSA, Municipalities and Local Authorities	59.966^x	61.3
State Economic Enterprises, and Medical Schools	17.255^{xx}	17.7
Social Insurance Institution	16.516	16.9
Private	4.028^{xxx}	4.1
TOTAL	97.765	100.0

Note: Health Institutions belonging to the Ministry of Defense are
      not shown.
  x   Only 2.645 beds belong to municipalities and local author-
      ities.
  xx 13.816 beds belong to medical schools
  xxx The figure include bed capacity of the hospitals operated
      primarily for profit and operated by voluntary and phi-
      lanthropic organizations.

Source: MHSA, General Directorate of Health Statistics and Pro-
        paganda.

Table 2 - Distribution of Physicians by Main Health Care Providers
           (1980)

Health Care Providers	No. of Physicians		TOTAL	
	Practitioner^x	Specialist	No.	%
MHSA, Municipalities and Local Authorities	3.685	2.881	6.566^{xx}	23.1
State Economic Enterprises, and Medical Schools	3.650	2.563	6.213^{xxx}	21.9
Social Insurance Institution	1.439	2.926	4.365	15.4
Private	1.970	9.297	11.267	39.6
TOTAL	10.744	17.667	28.411	100.0

  x Residents are included
  xx 605 Physicians were employed by municipalities and local author-
     ities.
  xxx 1.318 physicians were employed by State Economic Enterprises

Source: MHSA, General Directorate of Health Statistics and Pro-
        paganda.

State Economic Enterprises and especially Medical Schools
are other providers of inpatient and ambulatory care. At present
there are 19 medical schools in Turkey. They employ 17.2 percent
of the total 28.411 physicians and own 14.1 percent of the total
97.765 hospital beds in the country (Table 1). Most of the
medical research is carried out in medical schools and almost all
of them have field training areas where comprehensive care is
provided.

The Social Insurance Institution is the third main provider of
health care. This scheme was started in 1946 and has been financed
by employers and workers. Insurances of all industrial workers are
compulsory. This scheme does not cover agricultural workers. The
Social Insurance Institution has established its own dispensaries
(without beds) and hospitals in many parts of the country. At the
end of 1981 there were 80 dispensaries, 74 hospitals and 2 sana-
toria (with a total of 18.690 hospital beds) belonging to this
institution (7). In 1965 a new comprehensive Social Insurance Law
replaced the old programmes which expanded coverage to include
dependents of insured workers. There were 2.228.439 insured workers
and about 9 millions dependents at the end of 1981 (7).

The fourth main provider of health care is the private sector.
Nearly 40 percent of all physicians in Turkey work privately. Most
of them (82.5 percent) are specialists. They mainly work in their
private offices in urban areas and set their own fees. 5.338 out
of 6.790 dentists(78.6 percent) also work privately.

The fifth and the last health care provider consists of
quacks, traditional birth attendants, medicine men, religious
healers, bone-setters etc. A study done in 1963 (6) showed that
there were about 85.000 of these professionals compared to 21.000
paramedical health personnel in the country. Their numbers are
probably much less now. Kinds of medical care provided by these
traditional health workers are preferred by some people who are
poor and live in relatively neglected areas. Although these health
care providers are officially illegal, they are more accessible
and acceptable than state health services for some rural settlers.

2.2 The Coordination within the Health Sector

Cooperation between these five components of the health system
is rather poor. They have different health service systems and
belong to different Ministries. The private sector is powerful
and almost completely independent. Traditional medicine is both
illegal and ignored. Medical schools cooperate to some degree with
the MHSA, and there are signs that this cooperation will increase.
Except Red Crescent, all the other voluntary organizations are
rather poor and do not contribute much to the provision of health
care.

2.3 The Coordination of the Health Sector with Other Sectors

Coordination of health-related activities with other sectors is not much better. Part of the preventive services, especially those having to do with environmental sanitation are the respon- sibilities of municipalities and local authorities. Municipal leaders, being elected, give priority to other fields of activity which get more votes than health related activities.

Services related to occupational health, water supply, the sewage system and housing are carried out by other Ministries. MHSA has only supervisory responsibilities in these fields. School health services are insufficient. On the other hand, in terms of health education, not enough use is being made of the many primary and secondary schools spread all over the country.

Nevertheless, some voluntary organizations partly financed by the Government are doing useful work.

III. HEALTH RESOURCES PRODUCTION

3.1 The Supply of Manpower

Undergraduate training of health personnel is undertaken by Universities, Academies and by the MHSA. Medical, dental and phar- macy students are trained at Universities and Academies. Table: 3 shows the total number of students attending at all medical, dental and pharmacy schools during the 1979-1980 academic year together with the total number of 1979 graduates.

Table 3 - Number of Students Attending to Medical, Dental and Pharmacy Schools (1979-1980) and Number of Graduates (1979) in Turkey (8)

Kinds of School	Total no. of Schools	Total no. of Students Attending (1979-80)	Total no. of Graduates 1979
Medical	19	15.564	1.559
Dental	8	3.634	567
Pharmacy	7	5.043	877

Most of the paramedical personnel (nurses, midwives, laboratory technicians, environmental technicians) are trained by the MHSA. In addition nearly all medical schools have their own nursing schools. University level nurses have been trained at universities. There are two university level schools for health administrators, the one is operated by the MHSA and the other by Hacettepe Uni-

versities. Table 4 shows the number of students attending at
these schools during 1979-1980 academic year and total number of
1979 graduates.

Table 4 - Number of Students Attending at Paramedical Schools
          (1979-80) and Number of Graduates (1979) in Turkey (8,9).

Name of Schools	Total no. of Schools	Total no. of Students Attending (1979-80)	Total no. of Graduates 1979
Nursing (University Level)	4	950	137
Health Administration (University Level)	2	224	32
Health Education (University Level)	1	377	39
Laboratory Technicians (University Level)	1	308	23
Physiotherapist (University Level)	1	188	42
Health Colleges (Lycee Level)	95[x]	18.479	2.762
Private Nursing (Lycee Level)	2	393	100
Health Schools (Secondary School Level)	14[xx]	1.776	194

 x Eighty of health colleges train female nurses; 6 of them train
   midwives; 4 of them train male nurses; 3 of them train laboratory technicians and 2 of them train environmental technicians
xx These schools were transformed to health colleges which provide
   midwifery training.

     Systematic manpower planning is not practised in Turkey and
therefore at present there is an acute shortage of physicians and
a surplus of pharmacists and dentists. Recently, steps were taken by
the Government towards standardizing higher education in the country.
However, it is much more difficult to make basic changes in their
curricula adapting them to the basic needs of the country.

     It also proved difficult to overcome maldistribution of
health manpower, and to correct the unbalanced specialization.
At the end of 1980, about 59 percent of the physicians (16.081
out of 27.241 physicians) were living in 3 big cities (Ankara

Istanbul and Izmir) where 21 percent of the total population
reside (10). Almost three quarters of physicians are specialists
and they usually work in cities and big towns. Compulsory service of two years for medical graduates and specialists was
initiated in 1981. It is hoped that this measure will alleviate
temporarily the shortage of physicians in remote areas. Shortage
of public health physicians is a chronic problem. It is mainly
due to relatively low salaries and discredit attached to administrative tasks and preventive medicine.

The curricula of paramedical schools are more appropriate
to the health needs and their distribution throughout the country
is more even. Except nurses and laboratory technicians, there is
no great shortage of paramedical personnel.

3.2 The Coordination of Facilities, Equipment and Supplies

The pharmaceutical industry in Turkey is private. Most of
the substances used for medicines are imported. The prices are
set by the MHSA. The Ministry also allows license for production
of medicines and controls their qualities. Bacteriological vaccines
and sera are produced at Central Institute of Public Health, an
MHSA institution. Viral vaccines are imported.

Apart from those used in health units (Nationalized health
services) equipment and supplies are not standardized. This creates
problems of maintenance and repair. Lately initiatives for standardization have been taken by the MHSA.

3.3 The Coordination of the Supply of Knowledge or Technology

Knowledge and technology are produced by Universities as
well as by two Institutes attached to the MHSA, namely the Central
Institute of Public Health and the School of Public Health. Some
highly complex and inappropriate technology has been transferred
from highly developed countries causing problems of adaptation. At
times the MHSA makes contracts with or requests from the Universities to carry out applied research.

IV. ECONOMIC SUPPORT OF HEALTH SERVICES SYSTEMS

Over the last few years the percentage of funds allotted to the
MHSA from the Government budget has been decreasing. It was 4.2
percent in 1980, 3.6 percent in 1981 and 2.8 percent in 1982. The
budget of the MHSA shows a 22 percent decrease from the last year's
budget. If the 35 percent of annual inflation rate and about 25
percent of salary increase to the MHSA employees are also considered
the real decrease in the budget of the MHSA reaches higher
percentages. About 65 percent of the MHSA's budget was spent
annually for salaries and wages. Amongst the 16 General Directorates

of the MHSA, the General Directorate of Curative Services gets the biggest share which was 44.3 percent of the MHSA's budget in 1982.

To get a more comprehensive view of health expenditures, the money spent for health and medical care by Social Insurance Institutions, State Economic Enterprises, Medical Schools as well as costs related to private and traditional sectors should be added to the budget of the MHSA. These are very difficult to quantify and therefore were not included in this analysis.

According to a field survey carried out in Ankara in 1974, a middle class family spent one fifth of its monthly income for ambulatory medical care only (11). When expenditure for inpatient care is added, the figure reaches to surprising amounts. Preventive services like immunization are excluded since they are free of charge. This study also showed that in Ankara, the Capital of Turkey, one patient out of three seeks medical care from the private sector. Although they have to pay for the services, many people still prefer private physicians rather than government physicians.

Ambulatory care, apart from drugs, is free at health units (in Nationalized Health Services Areas) as well as at health centers and dispensaries (outside of Nationalized Health Services Areas). But payment is made for out-patient and inpatient care and surgical interventions both in state hospitals and in medical schools' hospitals for those who are not insured and able to pay. The government employees, insured workers and dependents of both groups obtain free medical care. It is estimated that about 65 percent of the population are not covered by any official health insurance scheme and have to pay from their own pockets for any curative care. Over the last few years a health insurance programme to cover all the population and financed by a special tax has been under discussion by top level decision makers but up to now no attempt has been made so far.

International cooperation mainly takes the form of project funding. The MHSA and medical schools receive financial support from International Health Organizations for research projects. UNICEF has also been helpful in provision of some medical equipments.

V. PATTERNS OF HEALTH SERVICES DELIVERY

Primary and secondary health care institutions in Turkey are summarized in Table: 5.

In Table: 6, specialized care institutions are given. As it is seen from Tables 5 and 6, not only primary health care but most of the secondary and specialized health care is provided by the MHSA.

Table 5 - Primary and Secondary Health Care Institutions in
         Turkey (10), (1980)

Types of Health Institutions	Owners of These Institutions	Total Number	Bed Capacity
A-Primary Care:			
Office of Government Physician[x]	MHSA	242	-
Health Centers[x]	MHSA	291	3.760
Health Units[xx]	MHSA	1.827	-
MCH Centers	MHSA	62	-
MCH Dispensaries	MHSA	65	-
MCH Statistics	MHSA	445	-
Dispensaries[xxx]	MHSA	326	-
Examination and Treatment Centers	MHSA	20	100
Dispensaries	Social Insurance Institutions	80	-
Private Physicians	-	-	-
Traditional Health Workers	-	-	-
B-Secondary Care:			
General Hospital	MHSA	236	33.446
General Hospital	State Economic Enterprises and Medical Schools	44	15.745
General Hospital	Municipality	5	1.549
General Hospital	Social Insurance Institutions	65	15.270
General Hospital	Private	70	2.497

x Function in non-nationalized area
xx Function in nationalized area
xxx Include dispensaries for tuberculosis, syphilis, leprosy
    and trachoma.

Table 6 - Specialized Health Care Institutions in Turkey (10)
         (1980)

Types of Health Institutions	Owners of These Institutions	Total Number	Bed Capacity
Maternity Hospital	MHSA	31	4.175
Maternity Hospital	State Economic Enterprises	1	12
Maternity Hospital	Municipality	2	870
Maternity Hospital	Social Insurance Institutions	3	735
Maternity Hospital	Private	12	275
Mental Hospital	MHSA	12	7.026
Mental Hospital	Medical School	1	120
Mental Hospital	Social Insurance Institutions	2	481
Mental Hospital	Private	5	935
Chest Diseases Hosp.	MHSA	30	7.965
Chest Diseases Hosp.	State Economic Enterprises	4	1.349
Chest Diseases Hosp.	Social Insurance Institutions	3	1.922
Chest Diseases Hosp.	Private	2	141
Oncology Hospital	MHSA	1	330
Oncology Hospital	Private	1	20

Collaboration between general practitioners and specialists is not satisfactory. A patient can get to be examined at a hospital polyclinic without first being examined and referred by a general practitioner. Specialists in hospitals are rather reluctant to give information to the general practitioner after consultation or discharge. Continuity of care depends on the patient alone. Referral from general hospitals to specialized ones work better.

Of all the groups in a community, mothers and children get the most attention. Activities of the health units are centered around the mother and child, especially those at risk. MCH centers are mainly located at towns and support the preventive and curative work of health units.

The Turkish population is a young one and therefore there is not much need for geriatric care yet. However there is an estimated population of 4.5 million handicapped including mental disorders. Services for these groups are provided by the MHSA and some voluntary organisations but supply is much less than demand. Although it is increasing, drug addiction is not yet an important problem in Turkey compared to some other countries.

Medical rehabilitation is the responsibility of the MHSA while occupational and social rehabilitation are almost non-existent. Thankfully the family in Turkey is traditionally a very supportive and tolerant institution and most of the care needed for the handicapped, mentally retarded or those requiring rehabilitation is given at home rather than at state institutions.

Recently, a reorganisation in the MHSA was attempted. Vertical services such as the tuberculosis and malaria control are being integrated into other services. The number of General Directorates is being lowered. MCH and Family Planning services are getting combined. This integration process in the Ministry is to be followed by integration in the provinces. It is hoped that all this reorganisation will increase both vertical and horizontal coordination within the health services.

VI. HEALTH SERVICES SYSTEM MANAGEMENT

Almost all of the MHSA administrators and provincial health directors are physicians. Few of them have had training in public health or management. Neither do the physicians in health units gain any skills in management during their medical education. This lack of managerial skills at all levels of the health system creates problems in planning, implementation and evaluation. Modern management techniques and the systems approach are hardly ever practised. Hospital administrators get university level training at two schools.

National Development Plans have a section on the health
sector. The plans are prepared by the State Planning Organization
with contributions from Universities and the MHSA administrators.
The Department of Planning and Coordination of the MHSA makes the
plans of the Ministry according to the National Development Plan.
National Plans are not obligatory to follow; they are mainly in
the form of recommendations. This usually results in their being
disregarded. Moreover, political parties may have different
priorities and policies than those set in the Development Plans.
Thus, plans may be altered with each new government. Therefore,
while the plans for nationalization of the health services have
always lagged behind, targets for hospital beds were overreached.

Budgetting is done by the MHSA at the end of each year and
the budget is presented to Parliament (now the Advisory Council)
for approval. After discussions on different items of the budget,
it is finalized and opened to the use of the MHSA. Both budgetting
and distribution of funds is done centrally without involving the
provinces in the planning process.

Health Policy decisions are taken by the Minister himself.
An Advisory Committee is available. Monitoring and evaluation of
the service is mainly performed through data forms received from
the provinces. Computer services do not exist. Service data and
notifications of communicable diseases, morbidity and mortality
are compiled by the General Directorate of Medical Statistics
and Propaganda and published as statistical year books.

In the provinces, personnel is supervised by the supervisory
staff of the provincial health directorates. Auxiliary nurse-
midwives are supervised by the public health nurse in the health
unit. Information systems and supervision work much better in the
provinces where health services are nationalized than those that
are not. However, notification of communicable diseases and mechanisms for feed-back of information leave much to be desired in
either system.

Some feed-back is provided by in-service training programs
held in the provinces for paramedical staff and also in the School
of Public Health for the directors of health of the provinces.
Physicians appointed to work in health units pass through a two
month adaptation course before they start to work. Furthermore,
the MHSA publishes two monthly magazines which are sent freely
and regularly to all health establishments in the country.

Rules, regulations, standards and norms are set by the MHSA.
Legislation, like the budget, is proposed by the MHSA and endorsed
by Parliament. Most of the health legislation is rather outdated
and requires revisions. We are well aware, however, that our problems related to the health system can not be solved by legislation alone.

References

1. State Institute of Statistics : Census of Population 1980
   (Summary tables); Publication no: 945, Ankara, 1981, p. 2
2. State Institute of Statistics : Statistical Yearbook of
   Turkey 1981, Publication no: 960, Ankara, 1981, pp. 399 -
   401
3. World Bank: Health Sector Policy Paper, February 1980, p.
   69
4. Hacettepe Institute of Population Studies: Turkish Fertili-
   ty Survey 1978, Vol. 1, 1978, p. 79
5. Ministry of Health and Social Assistance : General Public
   Health Law and Other Laws Which Amend the Articles of this
   Law, Publication no:178, Ankara, 1954, p. 3
6. Taylor, C.E., Dirican, R., Deuschle, K.W.: Health Manpower
   Planning in Turkey, The Johns Hopkins Press, Baltimore,
   Maryland, 1968, p. 177
7. Social Insurance Institution : Annual Statistics 1981,
   Publication no: 362, Ankara, 1982, p. 91
8. Devlet Istatistik Enstitüsü : Milli Egitim Istatistikleri,
   Yüksek Ogretim 1979-80, Publication no: 967, Ankara, 1981,
   pp. 4-14
9. Devlet Istatistik Enstitüsü: Milli Egitim Istatistikleri,
   Mesleki ve Teknik Ögretim 1979-80, Publication no' 971,
   Ankara, 1982, pp. 19, 27, 69-71, 78.
10. State Institute of Statistics: Statistical Yearbook of
    Turkey 1981, Publication no: 960, Ankara, 1981, pp. 89
11. Eren, N.: Ankara Il Merkezinde Saglik Hizmetlerinin
    Planlanmasi için Veri Toplama Yöntemi Geliştirilmesi,
    Doçentlik Tezi, Hacettepe Üniv. Toplum Hekimligi Enstitüsü
    (mimeographed), Ankara, 1974.

THE HEALTH SERVICES OF NORWAY:

ON THE ROAD TO A POLITICAL SYSTEM

H. Tore Skaug

Ministry of Social Affairs P. O. Box 8011

North Oslo 1, Norway

Introduction

Health services in Norway have for many years been a public concern and responsibility. The financing of the services has mainly been from public funds (approximately 95%). The responsibility for planning, construction and running has been with political boards on different levels (state, county councils and municipalities) and to some extend left to private entrepreneurs.

The health services institutions (hospitals, nursing homes etc.) have been the responsibility of the counties. The financing system has strongly stimulated expansion in their sector over the last decade. The counties have favoured somatic hospitals and nursing homes. Mental health services institutions and those for the mentally retarded have been lagging behind.

Responsibility for the development of primary care has been divided between central and local authorities and private entrepreneurs. No political body has had full responsibility for planning and running primary care. The result is a maldistribution of person-nel between urban and rural areas and an imbalance in personnel mix in the primary health services.

National policy, based on the regionalization of the health services, is to provide equal health services in all regions of the country (see chapter 0.1 below).

Two steps have been taken this year as a means to this goal :

1) The government has presented a draft law to parliament which will give municipalities total responsibility for planning and running primary health care within their boundaries.

2) A national health plan was presented in June, 1982.
The plan depicts lines of development with alternative solutions for different parts of the services within an economic growth rate of 1-3 % per annum.

If and when the Local Health and Social Care Act passes parliament and is made effective, the Norwegian health services will be a politically functioning system.

The fact that responsibility for the two main parts of the health services will be with two different political levels of government does not seem to be conducive to a coherent health services system.

This is the dilemma of Norwegian democracy and the decentralization of power. The central government has with parliament legislative and monetary/budgetting power and a main task in controlling and counselling local (county and municipal) authorities.

To fulfil the goals of the national health policy the central government is dependent on loyalty and solidarity from the two lower levels of government and from personnel organizations.

0. <u>Values, goals and objectives underlying the health services</u>

As mentioned above health services are a public responsibility:

a) to secure the availiability of health services to all inhabitants;

b) to cover the cost for the services so that they could be used by everyone.

More than 350 years ago, when the first positions for doctors were established in Norway, they were paid for by the national government on the assumption that most people were not in a position to pay for medical assistance out of their private resources. Medicine was thus "socialized" in Norway some centuries before the term was first coined.

In 1860 the Norwegian Health Act put the responsibility for environmental hygiene and control and the reporting of epidemic diseases on District Medical Officers engaged by the state in local districts. The posts for these medical officers were distri-

buted throughout the country. The District Medical Officer in most
communities combines the functions of public health medical officer
and general practitioner. This primary health care practitioner model is peculiar to Norway.

In 1930 a public insurance scheme was put into practice. It
provided the refunding of costs for hospitalized patients.

Hospitals, including cottage hospitals, nursing and maternity
homes, were mainly established by local entrepreneurs and industry,
by humanitarian/non-profit organizations and idealistic individuals,
and by local authorities with sufficient financial resources. Private practitioners (GPs and specialists) other than the public medical officers were free to set up practice. Nursing services were connected to institutions and medical officers and to a minor degree
to private practitioners.

Norway is a small country with 4,1 million inhabitants.
There have been vast differences in average income between urban
and rural areas and different regions of the country. The northern
most part and some parts on the west coast were sparsely populated
and poor. The entrepreneurs set up practice or established their
institutions in prosperous or densely populated areas. The outskirts
would be without services except for the District Medical Officer.
After World War II various laws intended to regulate the health
services passed the Norwegian parliament (Stortinget). The main
issue was to secure a more evenly distributed health service through
out the country. This policy was supported by all parties in parliament.

The first Act, Public dental services Act (1949) was followed
by: Disablement Act (1958), Mental Health Protection Act (1961),
Hospitals Act (1969). All shared certain features.

These Acts put the responsibility for the development of the
(institutions) health services within each county on the <u>county
council</u>. The Central government (The Ministry of Health and Social
Affairs) is to approve or modify the development plan. These Acts
contain regulations on how the cost of running and establishing is
to be divided between the political levels (national, county and
municipality) and the private sector.

The government proposed in 1971 in a parliamentary report
(No. 85 (1971-72) Health services outside institutions) that the
county councils also were to be responsible for planning and running primary care. The main purpose was to put all responsibility
for development of the health services on one political authority.
The time was not thought ripe for the "socialization" of private
practice so the issue was not pursued.

The government presented in 1974 a parliamentary report (No. 9
(1974/75)) "Hospital development in a regionalised health service
system". This report depicts the main principles and practical de-
lineation and norms for the development of the future health services.

The basic regionalization principle was characterized by an
organisational integration of all parts of the health services and
a functional, administrative and geographical definition of each
single unit. Thus each unit, i.e. doctor, home nurse, hospital etc.
is to offer certain functions to a geographically defined population.

The principle expressed the desire to stipulate a uniform
structure and common organizational framework of all health services.

The regionalization proposed would be carried out by a subdi-
vision of the country into 5 health regions, restructured into 3
main levels of health services :

1) District health services, comprising health services which
   ought to be planned on the needs of the local community

2) County health services - comprising health services planned
   on the basis of the needs within a county.

3) Regional health services - comprising types of services
   rationally planned for several counties (or regions of the
   country).

One of the aims of regionalising the health services was that
each patient could be treated at the lowest level at which the complaint can effectively be treated (the LEON principle).

The norms (and political goals) for various types of institutions
were as follows :

Somatic hospitals :   reduction from 5.4 tot 4.5 beds per 1000
                                 inh. (2.55 + 1.52 on county level, 0.43
                                 on regional level)

Psychiatric units :   in somatic hospitals : 0.2 beds per
                                 1000 inh.

Psychiatric hospitals :   reduction (from 2.0 beds per 1000 inh)
                                 followed by establishing of clinics
                                 and local nursing homes

Central institutions   :   for mentally retarded : same goal as
                                 for the psychiatric hospitals

Somatic nursing homes :	7.5 beds per 100 persons in the age-group 70 yrs. and older
Psychiatric nursing homes :	1,5 beds per 1000 inh.
Local institutions for mentally retarded :	No fixed amount, expansion following reduction of the central institution.

Cottage hospitals and maternity homes were to be retained in areas where the travelling time to nearest hospital was more than 2 hrs.

The Parliamentary report underlined the development of local health services and out-patient (policlinic) treatment as the means to reduce pressure on hospitals' intramural activities.

There was also a description of the future content of a local health service in the report. The service was to consist of : general practitioners, home nurses, school health services and mother and child care. This part was less binding since no political board was responsible for its development.

0.1 Basic political configuration of the nation.

Norway (4,1 mill. inh.) is devided into 19 counties (county commune) and 454 local municipalities (primary commune). Both the counties and the municipalities have their elected political boards and an administration to comply with responsibilities according to various laws (schools, health and social services, roads, land use etc.) They receive their funds from direct income tax, revenue on property and direct grants from the state more or less restricted to special purposes.

The population is homogenous even though there has been some immigration the last decade. Today there are approximately 80.000 (2%) foreigners in Norway.

1. Health services system structure

The health services in Norway can today only partly be seen as a political system governed by political boards on various levels.

1.1 Co-ordination within the health sector

By the introduction of the Hospitals Act in 1969 regulations concerning planning, construction, running, and financing of health institutions were changed and made homologous in the Disablement Act and the Mental Health Protection Act.

The counties are now responsible for somatic hospitals, including cottage hospitals, maternity homes, nursing homes and patients' hostels, psychiatric hospitals, clinics, nursing homes etc.,
private as well as public. The county council is each year given a
block grant from the National Health Insurance to meet expenditure
on the running of the institutions. The county councils drew up
their plans in accordance with the three Acts in the years 1970-73.

The county health plans are ideally dealt with in three stages :

I. Structure plan. This is the formal county health plan in
which it is decided where the institutions are to be situated, what
functions they are supposed to serve for a geographically defined
population. Formal approval is with the ' King in council '.

II. Construction plan. Physical building plan for hospitals are
approved at national level by the Director-general of the Health
Services, and for nursing homes and other local/small institutions
by the state's County medical officer.

III. Operational plan. Since 1980 a new rotating planning system has been in operation. The counties send plans for the development of their health services institutions to the Ministry of Health
and Social Affairs. The plan is supposed to be within the frame of
the structure plan. It covers a 4-year period of which the ministry
approves for the first year, and recommends lines of development
for the following three years. This plan provides a direct control
in the establishment of new posts for doctors, both geographically
and by type of speciality.

This planning system gives the opportunity of a neat control
of the development and coordination (both horizontally and vertically) of the health care institutions. (Financing will be dealt with
in section 3.).

The government apply the regionalization principle and the
standards mentioned in the approval of the county health plans.

The health services outside the institutions have so far not been
subject to the same sort of regulation (dental service for the
age-group 0-18 yrs is an exception). The responsibility for planning
and running primary care is partly with the state: district
medical officers , the county: local (commune) nursing officers
(child and mothercare), the commune: home nursing, School health
care etc., and private initiative : doctors, dentists, physiotherapists. The financing systems vary accordingly.

The establishment of new posts for doctors is regulated by the
Ministry of Health under the Intermediate Act on regulating surgeons
services (1979). But there is no means today by which the central

government can stimulate and co-ordinate the various levels of services.

1.2 Intersectoral co-ordination

The state and counties have clear and direct responsibility for sectors associated with health problems. Nutrition policy is one field in which one can see direct influence on people's health. This is taken into consideration by various state departments (fishing, agriculture, consumers (negotiations for sugar prices) and health and social services).

In other fields such as housing and transport policy, methodological problems frustrate full inter-sectoral co-ordination especially with regard to those sectors whose bureaucraties are otherwise only involved to a minor degree. The last decade has seen an immense improvement in co-operation on environmental pollution.

At the local level the district health services have both formal (through the local health board) and informal communication channels to solve both individual patient cases and planning problems.

Closed bureacratic structures hamper co-ordination also on this level. The Health Act (1960) and the Building Act (1964) compel the administration to follow certain procedures to secure hearings from different sectors before plans are put into practice.

2. Health resources production

2.1 Co-ordination of education

The Ministry of Culture and Science is responsible for planning and running of education at university level; the Ministry of Education covers the college levels; and the counties are directly responsible for the running of schools.

The planning system mentioned above, ad-hoc analysis supply and demand of various types of personnel and the long-term budgetting system should give a good basis for co-ordinating the supply of personnel. This has not been so up to this day. On the one hand the educational capacity at university level has been steered more by university policies than by supply and demand analyses. On the other hand there has been up till the late 1970's an expansion in budgets and in demand for services.

Public expenditure will need to be constrained during the coming years. The government will have the means through direct planning (at University level) and refunding systems to influence the capacity in the education system at the county level.

The problems will call more for political courage than for technical and strategic means.

3. Economic support of health services in Norway

The Norwegian health services are as mentioned above mainly provided from public funds. We spend nearly 20 billion Nkr. (3.1 billion US dollars) on the health services each year. This equals 8.0 % of the GNP.

3.1 Cost Control mechanisms

The National Health Insurance (established 1967) covers between 45 and 50% of the total cost of health services. The budget for the fund is presented to the parliament each year. The budget comprises the division of the expenditure between institutions and primary health care.

Preparation of the running cost budget is a part of the counties' implementation of the planning system for the institutions.

3.2 Financing

The insurance fund is financed by each working inhabitant paying 4,4% per year of his/her net income (school children and unemployed are exempt). The employers pay a varying sum of 8,6-16,8% of the employees income to the fund.

The funds's deficit is subsidised by the state budget.

The counties since 1st of January 1980 are given a block grant from the Insurance fund. The amount covers approximately 50% of running costs distributed between the counties according to objective criteria (such as total population, age and geographic destribution within the country, proportion of the population pensioned on grounds of illness).
The communes levy taxes on the wages and the property of the inhabitants.

The cost of the public dentistry (for the age group 0-18 years and to some extent elderly, handicapped and longterm institution patients) is divided on a near 50-50 basis between the state and the county.

Consultations by doctors and physiotherapists are partly covered by the Insurance fund (by a complicated and volumininous tariff system) and varying fees paid directly by the patient.

There is no private health insurance fund internally operating in Norway.

4. Patterns of health services delivery

Health services are based on the regionalization of the services depicted above. The services are constructed on a general basis, all integrated or amalgamated into the system. In this way we try to avoid duplicated provision or supply of care outside the regular services. The whole line of health services from the promotion of health, preventive, diagnostic, therapeutic and rehabilitation health care are public tasks and responsibilities (private practice by doctors, dentists and physiotherapists, is not forgotten, but also not mentioned).

4.1 Promotion of health

As starting point, promotion of health is or ought to be found outside the health services. Factors like environmental strain/ pollution, housing, recreational activities and nutrition are crucial to good health.

The central and local authorities co-operate in improving the physical environment. The central government distribute grants to the communes for recreation facilities such as gymnastics fields and halls, swimming facilities etc. Most of these are publicly owned. On local level the commune have their own officers for athletics and cultural activities. There is an established close co-operation between athletics associations and commune officials.

4.1.2 Preventive health care

Ideally all health workers ought to take part in preventive health care. This applies to general practitioners as well as specialists and other health workers. Preventive health care is built into the school system both as school health care (medical and dental) with screening and vaccination programmes and in the curriculum.

There are two state agencies with main tasks in preventive health care : the national council on smoking and health, the national council on Nutrition.

4.1.3 Diagnostic and therapeutic health care

The treatment of the various types of diseases is organized around the medical specialities and subspecialities of which there are 42. Specialized medical care is mainly given in the hospitals.

The hospitals are closed in the sense that the patient must consult a general practitioner to enter a hospital.

4.2 Function, place and organization of the primary care system

Primary care consists of the following elements: health board services and maternal child health care, school health services, general practitioners services, physiotherapy, public nursing duties (services) and home nursing.

The planning and running of these elements will, as mentioned above in the future be the responsibility of the municipality.

The health services within industry is today the responsibility of the industry itself. This will be integrated into the planning of the commune.

It is reckoned that primary care covers approximately 90% of all skilled health care. The rest is dealt with by specialized health care institutions.

The GP's are either part-time public medical officers or private practitioners. There has been a change from solo to group practices for doctors. In the last decade there have been established many local health centres where all the primary personnel have their offices. These buildings are in most cases constructed for this purpose and owned by the commune. With the health workers, even though their employers are the state, county or commune are obliged to cooperate.

The specialized institutional services operate either as consultants for the primary care or have special ambulatory services at health centers where these are found neccesary.

There is no political board directly responsible for the cooperation. This depends on local initiative.

5. Health services system management

The health services system comprises two main subsystems :

a) the structure and functioning of clinical treatment

b) the administrative//bureaucratic structure and functioning outside/by or on top of the clinical services trying to implement or make operational goals for the working of the services.

The first one consist of health personnel, the second of administrators and politicians. Some work in both subsystems.

In the clinical field entrepreneurship has been central and local autonomy essential for its functioning. This seems to be true both for specialist and generalist health workers.

To administer neccesitates simplifications. Most clinical health workers do not accept this for their field.

The administrators on all levels (at local political boards and in institutions up to central government) are in many ways "foreigners" to the clinical system.

The administrator trying to implement the general policy of distribution of limited resources is in many ways the antagonist of the clinical world.

Despite this the general political tendency and actual and concrete politics have led to solutions where one not only can treat the health services as a system theoretically, but systematically influence isolated parts of the system to make the whole work better.

To Norwegian health services factors like political responsibility for planning, construction and running are neccesary to gain the level of an operational system or political health system.

The above mentioned National Health plan will on the one hand make it easier for the politicians on all levels to take a stand and on the other give the health workers some background information for some unpopular measures that must be taken the next few years to keep the ship from sinking.

THE CANADIAN HEALTH SERVICES SYSTEM

Anne Crichton

University of British Columbia

Vancouver, Canada

0.0. Introduction

While a systems analysis of organisation design might be accepted by many Canadians, the concept of controlling the 'proper functioning' of a health services system 'within the limits of available resources' is much more controversial. The purpose of this institute "to examine steps taken to improve the 'coherency' of the different parts composing the 'whole' of the health services" (with implications of further development of bureaucratic controls) would be challenged both by the radical right and by the radical left in Canada.

The ideology of this approach is accepted, however, as the basis for the analysis which follows, but a large section of the paper will be concerned with an exposition of Canadian values and their effect on objective-setting and implementation.

The concept of the 'whole of the health services' is difficult to address since "health" is not defined in the outline. Canada is the least developed of the NATO countries and may be at a different stage of economic and social evolution for making good comparisons.[1] Although it has the second largest land mass of any nation, being 5514 km. east to west (with 4 time zones) and 4634 km. north to south, most of the population is now living within 150 km. of the southern border in metropolitan areas. Nevertheless, it is expected that health services will be provided to all residents. Therefore, the problems of health service delivery over vast distances and difficult terrain in a harsh climate must not be underestimated.

And, while the maritime provinces are relatively small and poor today, the provinces west from Ontario are large, relatively rich, and still in first stages of resource exploration and exploitation. Development of good road and air communications is quite recent. It is only since the mid sixties that Canadians have begun to look east and west, rather than south, over the U.S. border.

In 1871, just after Confederation, Canada had a population of 3.1 million, mostly living in rural communities. In the forties (when the main health service legislation was worked out) it had 11 million, now it has 24 million.
The assimilation of a very large number of immigrants since World War II into an evolving nation has posed problems relating to integration. A bilingual, bicultural policy (English/French) of the sixties now has a multicultural policy superimposed in the seventies. This still does not satisfy indigeneous peoples who are concerned about their economic rights as well as their cultural identity. If we take the World Health Organization definition of health* as the objective, it is clear that Indian land claims have become a more important health matter than providing, say, a nurse practitioner service. Since their first court case over land claims was won in the early 70's, the native peoples are regrouping. While they still have a higher incidence of alcoholism, infection problems and violent episodes requiring medical treatment than other populations, this is changing as their confidence grows.

In general use, however, "health services" are those services where gatekeeping for society is done by physicians. Sometimes they are seen as an end in themselves sometimes as a means to other ends; but services set up for other reasons entirely may ultimately be measured in terms of their contribution to health status improvement and charges to Ministries of Health.

0.1. Values, Goals and Objectives

The values underlying the health services system in Canada are ambiguous. As Alford[2] has indicated, conflicts in values may lead to 'dynamics without change'.
Yet there has been a considerable amount of change in the last forty years because there was agreement in the forties between all interest groups to proceed towards collectivist policies and a series of federal Acts was passed establishing a national health insurance scheme.

---

* "Health is a complete state of physical, mental and social wellbeing and not merely the absence of disease or disability".

In the sixties the underlying conflicts between health
professions and governments began to emerge again. These
conflicts were smoothed over by providing strong financial
incentives and making other concessions to the providers of
services, but they have begun to resurface in these inflationary
times.

The agreement of the different interest groups (which,
using Alford's terminology we can identify as entrepreneurs,
corporate planners and consumer groups) to a collectivist
policy came at a particular stage in Canada's evolution. It
was a product of the thirties when a poor, underdeveloped
country had problems in providing adequate social services to
citizens in need and of the war, when rewarding veterans
equally for their service became an issue of ensuring minimum
standards and more equitable distribution. But, in the post
war years, Canada's economic development 'took off' and frontier
attitudes of entrepreneurship and self-help have gained ground.
As in the rest of North America, the nearer one gets to the
frontier the more dominant become self-help attitudes, with a
'residual welfare'[3] approach to income sharing. As a result,
such redistribution of services as exists is likely to be defined
as health care rather than social welfare, because of the stigma
associated with the latter.

There is an ambivalence or an inconsistency in the Canadian
public's values to-day. Few, if any, would wish to revert to
pre-national health insurance days, yet there is the frequently
voiced suspicion of the many "small l" liberals that government
involvement should be restricted[4]. Spokesmen for "entrepreneurs"
continually attack government planning activities unless these
happen to coincide with their particular interests. If, as
Lowi[5] suggests, governments evolve through distribution (pork
barrelling), and regulating, to redistributing, Canadian
governments are still in the early stages of evolution. Nor
have they accepted the concept of "limits to growth"[6] because,
so far, their experience of resource development is that it
has provided more opportunities and more wealth for
redistribution.

In this atmosphere of frontier entrepreneurialism what of
medical entrepreneurs?
The most extremist supporters of medical free enterprise seem to
have been drawn off into the U.S.. As described below in
Coordination of facilities, equipment and supplies, Canadian
profit making activities in health care are relatively minor
though there is still an attitude of mind in the medical
profession which runs counter to collectivist solutions. This
is partly to do with physicians' payment on a free-for-service

basis (which is strongly defended) but is also partly to do with
the value set on scientific advance. There is, at the heart of
all western developed nations' health care systems, a conflict
between support given to development of high technology or
to the low technology of improved community care. Canadian
physicians and hospitals tend to favour the former. Although
they are not 'the willing tools of big business' (as some
critics[7] have described their American counterparts), they resist
government controls over resource allocation and redistributive
solutions. Johnson[8] has argued that a profession's relationship
to others may be described as following one of three models of
power - patronage, collegial control or mediation. In the
present technological age, the mediation may be that of the
health industry or of government. Canadian physicians are torn
between the two (as Lang[9] has shown in his discussion of the
control of pharmaceutical distribution in Canada) but since
there is no strong Canadian health industry outside government
and its delegated authorities, the incentives to profit-making
through exploiting advanced technology are not so great as over
the border. The situation of the medical profession is, rather,
one of defending arms' length negotiating mechanisms with
governments, the bursars, of maintaining the position of being
subsidized entrepreneurs rather than being in contract with
governments, of not being hospital employees but having hospital
privileges as independent contractors, in order to keep a wider
ring of autonomy round clinical decision-making activities.

There have been considerable stresses and strains in
government-physician negotiations. The collectivist agreement
of the forties was soon overtaken by the entrepreneurial
ethos espoused particularly by immigrant refugees from the
National Health Service and by the early sixties the doctors
in Saskatchewan went on strike[10] against the takeover by government,
as third party, collecting contributory prepayments for medical
services and paying their fees. Although it was agreed to go
ahead with such a development on a national basis after further
investigation by a federal Royal Commission (1964)[11], the
ambiguities in the relationship were not clarified. They have
now resurfaced and remain unsettled because some governments
themselves are not totally committed to collectivist solutions.

Because of these liberal attitudes to minimal government
involvement, one cannot assume that all governments of Canada
are strongly committed to corporate planning solutions. In fact,
the growth of government bureaucracies is a recent phenomenon,
partly because of this concept of 'the least government the
best government', partly because provincial governments had
few resources to develop bureaucracies until the economy began
to be developed simultaneously with the offer of federal matching
grants for social services. The larger bureaucracies set up

in the fifties were (still are?) clerical assembly lines rather
than professional cadres (see 5.1). Corporate planning as an
activity was promoted by federal civil servants and some
academics but is barely beginning to be accepted at provincial
levels[12]. The political bureaucratic frontiers are still being
drawn and redrawn in the smaller jurisdictions and this deters
many senior professional health administrators from entering
or remaining in government service at the provincial level.
Consumers of health care in Canada are not a powerful, stable
pressure group. They are not well organized into strong advocacy
groups and their special interests tend to splinter their efforts,
e.g. between supporting development of services for a particular
disease or against particular treatment methods or acting as
board members of an institution. The closer they get to sources
of power for lobbying purposes, the more likely they are to be
coopted and used as buffer groups by these stronger powers.

There has been significant change in the position of
individual consumers, however, in the last decade. A Health
Charter for Canadians, incorporated in the Report of the Royal
Commission on Health Services, 1964[11], laid out the rights and
responsibility of consumers. For the first ten years thereafter
the main emphasis continued to be laid on rights as federal
government moved into the end stages of its collectivist national
health insurance programming. In 1974, a report of the Long
Range Planning Branch of the Department of National Health &
Welfare[12] reviewed the statistical evidence about life-expectancy
in Canada and emphasized the need to reconsider existing health
policies with the sole emphasis in policy making upon health
care organization and funding. The "health field concept" put
forward in that report argued that four sectors should be of
concern: lifestyle, environment and high risk, in addition to
"A New Perspective on the Health of Canadians", demonstrated
consumers' positive feeling for this restatement of values
regarding the professional-patient dependency relationship -
reasserting greater personal responsibility for health (though
some have seen this policy shift as a sinister move towards
blaming the victim).[14]

The Health Charter for Canadians did more than address health
needs, however. The Commission made it clear that development
of a national health insurance scheme was an appropriate mechanism
for integrating Canadians into a nation. This was not the first
time that health services had been seen as a means to an end
rather than as ends in themselves. There is no doubt that the
first of the series of federal programs of matching grants
offered to provincial governments in 1948 was not developed
entirely on arguments for improving health status but as an income
redistribution, job and small business promotion scheme in addition
to providing the symbols of caring through hospital construction.

(Indeed, corporate planners had clearly recommended starting
with primary care if improved health status was the main
objective[15]). Nowadays, it is obvious that health service
development has served its turn in the promotion of Canada
as a national entity and energy has taken over. So, health
services are more likely to be evaluated in terms of their
health status achievements in future. As some analysts have
pointed out, the further one moves away from examining the
program level towards the social policy development level,
the more likely it is that inconsistencies in setting
objectives or working towards goals will be found[16]. In Canada,
these inconsistencies arise out of its rapid evolution as a
developing nation and the ambiguities in values, objectives
and goals, as well as the complexity of jurisdictions to be
described below. The question 'Who is the client?' for the
health planner is still very uncertain. In the seventies,
epidemiologist planners began to focus attention on health
status indicators for their Ministers of Health but, with
the present cost pressures, financial experts have moved
in to develop other rationales for maintaining or cutting
programs, namely their immediate votegetting qualities for
Cabinets as a whole[17]. It is difficult to cut well established
schemes but new approaches to program evaluation are emphasizing
obvious value for money and the need for greater managerial
control to improve efficiency and effectiveness.

     To summarize, the main trends in the post war years: the
primary objective of establishing a federal collectivist
welfare state was achieved by 1970[1]. In the subsequent period
this achievement has been accepted as a given by citizens
generally, despite some questioning by involved professionals;
a second objective, namely improved management of the services,
was first stated in 1969 in the Report the Task Force on the Cost
of Health Services[18]. Integration of programs began to be
explored thereafter and is now the principal objective of
governments, although professional resistances to streamlining
are likely to increase.

0.2. Basic Political Configuration

     It must be recognized that although a separate Canadian
identity has been a major political issue since Confederation,
and particularly since the late 60's, Canada's dependence
on Britain is just about to be severed by new constitutional
arrangements. Particularly important is Canada's internal and
external economic dependence on the U.S..

A recession there has enormous implications for the Canadian
economy, which provides primary resources for U.S. manufacturing.
At the present time, this recession has begun to force the
Canadian provincial governments to examine their health and
social service commitments and to review cutbacks despite
basic richness in, as yet, untapped resources.

The patriation of Canada's constitution in 1982 will not be
a revolutionary but an evolutionary move. New legislation
will be built on the precedents set by the British North America
Act, 1867. Power is now divided between the federal government,
ten provincial governments and two territories. It was not
foreseen at the time of writing the Act that health and social
services would become a major responsibility of provincial
governments.

Federal government is responsible for defence, economic
development, manpower and immigration, care of native peoples
and social security payments to identified risk groups.
Under a constitutional revision of the forties, it makes grants
for specific purpose in health, education and welfare to
provincial governments. A Charter of Rights appended to the new
constitution guarantees free movement of people, goods and
services across Canada, just as federal legislation relating to
redistribution of federal taxes establishes basic conditions
for programs established with federal support.

There are great differences in economic resources and uneven
economic development in different provinces. These differences,
and political party differences between federal and provincial
levels, have led to struggles between federal and provincial
governments. The Fathers of Confederation did not foresee
that provincial governments would develop such bargaining power
as they now have. Led by Quebec (which bargained for block
grants instead of tied grants in 1964)[19], the provinces have
argued their constitutional rights and their superior knowledge
of local needs for education, health and welfare spending,
but Federal government, in allocating monies (en bloc since
1977 to all provinces) has insisted on maintenance of the basic
health and social security programs established from 1948-67
in federal legislation. There is some skirmishing around this
concept at the present time (see 3.1).

The adoption of redistributive policies has resulted in
an unintended shift of power from central to provincial
governments. Presently there is an imbalance between the
provinces' responsibilities to provide services and their
taxing powers.
Canada's federal and provincial governments conduct their
political business following the British Parliamentary model.

Federal-provincial conferences are the mechanism for negotiating boundary problems.

1.0. Health Services System Structure

Under the B.N.A. Act, provincial and territorial governments (with minor exceptions listed below) regulate and redistribute resources in support of health, education and welfare services within their own boundaries. The Federal government's responsibilities were discussed above.

The principal problem of the provincial government is that, because they were unable to deliver direct services themselves (even if they had wanted to), in the early years of confederation, they delegated all their responsibilities to municipalities, registered charitable organizations and entrepreneurial practitioner associations (save for providing mental health services to control those who could not be managed in their own communities). In the succeeding years, the delegated authorities entrenched themselves and the introduction of national health insurance programs tended to strengthen their positions. Now the shift in government's objectives towards coordination and improving efficiency and effectiveness challenges established organization structures. New structures will be difficult to develop.
Ambivalent feelings about the powers and purposes of governments have not resulted in strong governments. While Ontario and Quebec, as larger provinces, came to recognize the need for improved organization in the sixties, it was not until the mid seventies that British Columbia (B.C.), for example, took steps to implement a bureaucratic management model to back political decision making. Nevertheless, the decentralization of government power to other authorities has been reconsidered from time to time, usually when crises have arisen (such as failures in infection control, political dissention between provincial and municipal bodies about the handling of social assistance clients).

Municipal authorities of the 19th Century proved to be too small by WW II and were conjoined or replaced by regional organization or monitored by provincial inspectors. The rationales for modelling a regionalization of health services have varied from one province to another and in the actual implementation of such policies the original rationales may be set aside as tradeoffs occur.

Philanthropic charitable organizations continued to be important bodies till the mid sixties and they still play a part in expressing community concern and in pioneering new activities, but a review of voluntary organization and volunteerism in the seventies[20] indicated that they were falling behind their targets for a number of reasons(self-help developments, decreased availability of volunteers through women joining the labour force, increased taxation and redistribution by governments).

Entrepreneurial practitioners and their workshops, the hospitals, developed their own self assessment mechanisms for control of standards[21]. In Saskatchewan, these were backed by legislation. This does not seem to have been regarded as necessary elsewhere with the possibility of taking cases to court as a final sanction. (Litigation is growing[22] but is very minor as compared with the U.S.). This self evaluation ethos prevails and is now being used as an argument in collective bargaining by health workers at all levels (with negative consequences for any attempts to apply a control rather than a negotiation model of management).

1.1. Coordination within the Health Sector

Matching grants to the provinces (which were implemented) were offered in the following order:

1948 (a)   on the basis of a written provincial plan, hospital
           construction grants;
     (b)   demonstration and training grants for improvement
           of existing public and mental health services;
           (National Health Grants);
1957   grant to meet hospital operating costs; (Hospital
           Insurance and Diagnostic Services Act);
1965   grants for development of research and teaching
           facilities for health workers; (Health Resources
           Fund);
1967   payment of doctors' fees; (Medical Care Act).

As explained above, rational health planners would have differently proceeded[15] but the Federal government had to recognize other rationales and bow to political expediency. The long term goal was to legislate a national health insurance scheme in whatever order was possible, rather than "the correct order". In consequence, the integration of programs was a second level objective.

In 1948, there was insufficient bureaucratic expertise to prepare and implement good provincial hospitalization plans and local political forces soon took over.

It quickly became clear that insufficient controls over
hospital construction and strong incentives to hospitalize
patients accelerated demands for hospital operating funds.
Prepayment schemes for primary medical care were less
important to voters than crisis care in hospitals and so
were delayed in being developed.
Then there were struggles for control between medical
professionals and governments. When the Medical Care Act
was passed in 1967 a Task Force on the Costs of Health
Services[18] was called together. It reported in 1969 that
there was a need for improved integration and rebalancing
of services. Its main recommendations may be summarized as:

1) change the funding arrangements in order to change the
   financial incentives away from hospitalization of patients;
2) explore new forms of organization such as community
   health centres, regional structures;
3) increase outpatient care;
4) explore labour substitutions.

Federally funded investigations and demonstrations of the
70's which followed[23] were not received positively until the
"New Perspective"[13] report was published in 1974, and that
still needs to be developed from a value statement to an
effective strategy of implementation.

Meanwhile, many provincial governments awaiting
reorganization of funding arrangements followed Quebec's
example, developing critical reports on their health service
organization. These were mostly concerned with better use
of government funds - better control of delegated activities.

The issue "how clinicians ration services" is a cause
of concern to many[24] though it is not often discussed in those
terms and so not well understood. Physicians have themselves
(for reasons of maintaining autonomy) established monitoring
of economic activity on a comparative basis, but there are no
absolutes, only relative and a shifting norms of medical
practice.

Capital and operational funding of institutions is a
matter of political negotiation rather than of rational
planning or strategic management as yet, though there seems
to be more prospect of governments gaining control here in
times of economic recession. Capital development controls
have been relatively successful but operational funding is
still bargained out by individual hospitals. One mechanism
which has been used to control outflows of funds is the buffer
group with a limited budget - regional districts, collective
bargaining agencies, subsidized consumer groups (which

become dependent on the subsidies). But buffer groups can
become pressure groups exerting their power the other way
round too. The impact of "A New Perspective" is slow in coming
but it is likely to be as great an influence on policy development as the renegotiation of funding arrangements because of
its redirection of objectives over the long term. It will take
time to gain acceptance for greater emphasis on health promotion programs. Environmental issues require new coordination
mechanisms to be developed across existing boundaries, e.g.
with Workers' Compensation Boards, Motor Vehicle Branch
(re seat belts, drunk driving, etc.), but these are emerging.

1.2. Coordination of the Health Sector with other related sectors

There is a Department of National Health and Welfare which
shares the same Minister and the same building but there is
inevitable separation of activities down the hierarchy [25].

At the provincial level there is a spectrum. Quebec has
one Minister of Social Affairs [26] and in its model of primary
care, the Community Health and Social Services Centre, one
intake worker. In other provinces, there are separate Ministries
which provide services but these are structured differently
across the country. To take two examples, Alberta has a
Ministry of Community Health and Social Services whilst B.C.
has a Ministry of Health and a Ministry of Human Resources.
Ministries tend to be restructured in relation to one another
or within themselves with great regularity. Provincial governments are unwilling to give one Minister the whole of their
health and welfare budgets to control and so divisions must
be found.

The stigma attached to accepting help from social welfare
was explained in Basic Political Configuration. Some provinces
have overcome this, others must designate 'drop outs' as
stick [27].
In B.C. and Alberta, for example, it is not possible for the
elderly to get Long Term Care except through medical gatekeepers
while Saskatchewan and Manitoba permit social service admissions.

Inter-ministerial power struggles take place around health
manpower planning [27]. A particularly blatant example of competing rationales was the expansion of the Medical School in
British Columbia where the ratio of doctors to population is
1:468.6 [28]. The justification for the expansion, funded by the
Ministry of Education, was said to be that bright young people
in the province were not having equal chances to enter the medical profession as compared with other provinces, but the Minister
has also shown his real hand by promoting other advances in high

scientific technology, whilst the Deputy Minister of Health was
engaged on a campaign to promote low technology in health care.
Manpower planning, generally, is about thirty years behind
England and is further hindered by the Charter of Rights which
permits free migration within Canada.
Educational planning is a provincial matter whilst immigration is
a federal concern. In a country which has admitted people since
the war, educational planning is only just beginning now
immigration has been halted.

1.3 & 1.4. Federal, provincial, regional, district & institutional coordination

It is not proposed to discuss coordination much further
here. The feasibility of coordinating well depends upon the
legitimacy of the concept and support for implementation.
Despite hesitations, the value of coordination is beginning
to be accepted but feasibilities will require much working
through.

A major issue is the method of coordination - organizational
design or managerial control in what proportions?[29] At the
present time there is a lack of sophistication in both.
Economists and lawyers are exploring the question of
professional regulation as a central problem[30], since they believe
that the problems of coordination are unlikely to be resolved
without introducing changes restricting professional autonomy
to the immediate doctor-patient relationship, whether by changing
professionals' power in government funded organizations or by
returning to the medical marketplace.

2.0. Health Resources Production

2.1. Coordination of Education & Supply of Manpower

Responsibility for planning health manpower development
is unclear. There have been ad hoc and continuing studies of
medical manpower production and distribution since 1964 and
other health professional groups are being studied at the present
time[31]. Since most health workers are women their choices are
difficult to predict.

An open immigration policy and rapid development of post
secondary educational institutions since the mid sixties has
resulted in an oversupply of some professional groups in some
places and a shortage in others. Distribution of manpower
and coordination of planning were issues discussed in 1.2. The
facts about education, training, licensing and registration
are available but these tell us little about choices of individuals

to stay in, or move out of, the health labour market.
Planning of the mix of an institutional labour force
presently takes place at the institutional level. More planning
is now being done by provincial Ministry of Education
coordinators of health courses in conjunction with
institutions both in terms of vertical course progressions for an
occupational group (e.g. nurses) and for overall manning issues
(horizontal planning). Interprovincial and intraprovincial
coordination between Ministries and institutions in health
and education is also increasing. But to take one example,
B.C. decided to introduce a Long Term Care Program within
six months in January 1978, for political reasons.
There was no pre-planning of manpower needs. In consequence,
organizational design was less than optimal. Nurses used as
assessors did not have community care attitudes, there was
an inadequate supply of homemakers. The demands for insti
tutional care continue to rise for these and other reasons
(e.g. housing market problems)[32].

There has been considerable elasticity in the existing system
which has permitted adaptations (e.g. increased numbers of doctors
have been able to increase volume under fee-for-service), but
funds are getting tighter and governments are anxious to get
more control. They will have a lot to put together and the
mechanisms that have to do it are not yet properly developed.

2.2. Coordination of facilities, equipment and supplies

It was about twenty years before governments were prepared
to take a firm line on hospital construction planning after
the introduction of national health grants in 1948. Federal
government pulled out of construction matching grants in
1969[12] but by then provincial mechanisms for coordinating
facilities and equipment planning were in place. These are still
subject to violent political upheavals when governments change,
but otherwise governments control developments reasonably
well at the managerial level. Closing facilities is more
difficult to achieve.

The rationales for allocation of resources are still
uncertain. A model for developing a hierarchy of hospitals
for a large province with one medical school is being slowly
negotiated in B.C.[33] following similar processes in Ontario
and Quebec, but such threats to the existing referral system
cannot be dealt with quickly.

There is no large Canadian supplier of equipment and supplies.
National defence requires support of laboratories producing
plasma, vaccines, etc. and they are seeking out supplementary

markets abroad. On the other hand, Canada is colonized by
other countries' drug and equipment industries (particularly
by the U.S.). The federal government struggled to gain control
over pharmaceuticals in the sixties by developing national
guidelines on formulae, substitution of generic drugs for
named drugs, etc.[9] Policies have been worked out seperately
by each province on these matters. There is probably an
oversupply of pharmacists who have been rescued from hunger
by the introduction of provincial pharmacy support schemes
for the elderly (Pharmacare).

Control over laboratory tests varies from province to province.
In general, provincial government laboratories deal with
infectious diseases and difficult referrals from hospital and
private laboratories. The latter do the standardized tests.
Volume control is exerted by negotiation between governments,
pathologists and physicians who are made aware of practice norms.
Similarly, radiologists are expected to exert controls.

The continuing pressure to keep abreast of developments in
technology has already been mentioned.

2.3. Coordination of supply of "knowledge" or "technology"

Canada has a small community of scientists who are
cosmopolitans in close touch with national and international
developments[34]. In order to be able to draw from the knowledge
network, it is believed that Canadians should also be contributing
to it, thus grants for research are provided (self initiated/peer
reviewed and commissioned), facilities are funded by special
federal grants (1966-80), conference attendances and study
leaves supported so that information can be exchanged or
collected. Libraries in universities are excellent. Publication
is encouraged.

So far as grass roots practitioners are concerned, hospital
libraries are well stocked and continuing education is subsidized,
but there are problems of distance. Consultation by phone is
encouraged. The Health Sciences Centre concept was promoted in
Canada in the late sixties[35] but interdisciplinary learning in
universities did not catch on. There was, however, a greater
willingness to try teamwork in clinical settings. But experience
with the problem-oriented medical record shows a reversing of
this trend in the 1980's back to physician dominance[36].

3.0. Economic, including Financial, Support of Health Service Systems

Unlike other countries, Canada does not have much of a private

sector in health care, nor have Canadian governments yet tried
to promote competition from the private sector,though they may
well be on the brink of doing so as the demography of the
country changes. This does not mean that governments fund
everything. There is still some entrepreneurialism, still some
user charges. The entrepreneurialism takes the form of private
capital and operational funding of doctors' office
facilities, of endeavouring to negotiate "subsidized
entrepreneurialism" in the fee-for-service sector (see 0.1.),
of some provision of long term care "for profit".
Pharmaceutical prescriptions and dental work are likely to
be charged for unless there are government schemes for special
risk groups but there are employee insurance schemes for
additional benefits. User charges vary from province to
province[37]: extra billing by physicians may be permitted by
provincial governments (though federal government is
threatening to withdraw its subsidies if this practice
continues), per diem charges may be made for hospital care,
emergency department, or ambulance use may be billed.

3.1. Coordination of Efforts to Control Total Costs

The largest percentage increases in health service
costs came before the end of the sixties[38]. In the
seventies the Canadian governments have been relatively
successful in holding these costs at around 7% of G.N.P.

Following the Task Force on the Cost of Health Services[18]
and other investigations of the early 70's[23], the
Established Programs Fiscal Arrangements Act, 1977, was
negotiated in order to close off the original open-ended
tied grants of the federal health, education and welfare
legislation of the post-war period. The decision of the
federal government to move to a block granting system was
not readily accepted by the provinces and the agreement was
to be renegotiated after five years. When a Conservative
government took over in Ottawa in 1979, it came under
pressure. A Royal Commission was set up to report on
"Canada's National Provincial Health Program for the
Eighties".[38] The Commissioner, Hall, was to review the
publicly financed health insurance programs within a social
and economic climate which had significantly changed, to
consider basic principles and nature and extent of
necessary revisions.

He identified current problems as financial: "There
can be no doubt that these two programs (Hospital Insurance
and Medicare) served all parties well but there were a number
of factors that led to major changes in 1977". Hall said that

there were two separate, if related, problems:

"1.   The first was that by federal sharing in the costs
      of only two services, these being not only the most
      expensive, but also the programs having the highest
      rates of increasing costs, the effect was to create
      an imbalance in the overall health services system.
      That is, the conditional grants-in-aid had
      undesirable steering effects in Provincial
      decision-making.

2.   From the federal point of view the hazard in the
      system was the total lack of control over its expenditures.
      Each fiscal year the Federal Government had to
      provide 50% of whatever amount the Provinces
      collectively had decided to spend. The annual
      increases in the late sixties and early seventies
      were large, exceeding the increases in the
      Consumer Price Index in every year."

He concluded that the provinces were fulfilling the conditions
of providing the care they were expected to give but that the
economic situation had changed since the negotiations
had taken place in the mid seventies and so there was some
disappointment with the term of fiscal federalism.

A (federal) Parliamentary Task Force on Federal-Provincial
Fiscal Arrangements[39] set up to examine the problem reported
in 1981. Its conclusions were that "the achievement of
comprehensively funded, hospital, medical and extended health
care is a major accomplishment of Canadian society ... (which)
could be jeopardized by reductions in current aggregate levels
of federal support because such reductions would be likely
to lead to increasing reliance on private funding and ultimately
to higher health care costs and erosion of the program
principles". The Task Force recommended (1) renewal in its
present form, with at least five years guarantee (2)
annual reporting on provincial compliance, (3) sanctions
for non-compliance, (4) consolidation of legislation with
more explicit spelling out of criteria as a basis for
monitoring.

It is easier to control facilities and equipment costs
than operating costs. Despite end runs by high technology
specialists in teaching hospitals, the former is reasonably
well managed and is getting better managed.

Manpower issues were discussed in 1.0, 1.1 and 1.2.
Jurisdictional issues and differing rationales prevent good
control.

Operational funding can be divided into direct and delegated activities. Direct services under control of Treasury Boards and Public Service Commissions have been restructured into tighter organisations (or so it is hoped) and are being more closely managed than a few years ago. Delegated activities have to be negotiated but these negotiations are naively conducted. Coordination of organizations is only just beginning (e.g. Alberta study of cost mix, [40] B.C. Hospital Role Study, [33])and management has been lax. Deficit budgeting has been permitted (encouraged?) and this is only now beginning to be replaced by global or zero based budgets. The success of these is yet to be established. Occasionally, however, lines are drawn. In B.C., for example, many long term care "for profit" institutions were forces to comply with government guidelines on changes. [41]

Personnel (doctors, nurses, other unionized staff) are more sophisticated at negotiation than their opponents. There is a lack of clarity about the limits of negotiation of terms and conditions of employment and wage settlements of 40% over two years are not uncommon.

3.2. Coordination of Sources

The predominance of government funding in Canada has been described above (0.2, 1.0). The physicians did not wish to enter the Medical Care program unless it was contributory [42],so employers in some provinces deduct from employees or pay their contributions as part of a bargaining package.* Contributions are a minor matter in health service funding however. Employers also fund Workers' Compensation Boards and may provide occupational health services in their plants. Unions are now involved only in getting better wage bargains, though they had contracted doctors in the past [43].

Organized voluntary agencies for health and social services and other local community efforts tend now to be program gap filling and demonstration activities, or to show concern that is not purely financial but also compassionate. [44] Attitudes are altruistic in many sectors which might be commercial elsewhere; thus the Red Cross

----

* When residents are unable to pay their contributions such governments provide help to ensure that their coverage is maintained.

collects and processes blood, provides equipment for
hire and undertakes international missions.

There are, of course, special interest groups which
try to get more of the pie . On the whole, the general
public is not very discriminating in its giving, spurred
on by T.V. shows or good publicity campaigns. From time
to time there are enquiries into commercial promotions
and their rake-off. Individual contributions to
registered charities are income tax deductible.

Private households are required to pay for services
not covered by the health insurance scheme. Some examples
of user charges were given above [37] (3.1). In recent years
governments have offered more and more gap filling programs
to special risk groups, in such areas as pharmacy support,
denticare, Aids to Daily Living and extended care for the
elderly and handicapped. [45] Many employed persons and their
families have extended benefits plans negotiated with
employers to meet user charges for pharmacy and dental
care and for the fringe areas around acute care (e.g.
private rooms in hospitals). The complaint among non-
metropolitan dwellers is not usually about health care
costs but about costs of transportation to get access to
consultants - a question of equity in distribution of care.
[46] These costs can be very high but programs to regionalize
treatment services for cancer or arthritis, for example,
have helped to shift the issue to complaints about elective
consultations.

The Canadian federal government has a series of international agencies which are active in the health field.
For example, C.I.D.A. is expected to promote the use of
Canadian products abroad and to maintain a Canadian presence
in countries with important sociological or economic links
to Canada. This is important for Canadian resource production
(other than building of facilities). Such production is concentrated in a few laboratories (see 2.2).

4.0 Patterns of Health Service Delivery

The trade-offs required for reaching agreement to
develop the Canadian health system resulted in an un-
economic emphasis on acute care. Following other countries'
examples, the federal government is endeavouring to turn
the system round towards low technology and towards
applications of existing knowledge rather than

creation of new knowledge (as from 1969).
Establishment of a new Health Care Research Fund[12]
indicated this shift. Response to "A New Perspective"[13]
established public support at least for
individual efforts to be healthy.

4.1. Coordination of Promotion of Health

Canada has always had a strong public health service
based on the British model of sanitary control and
personal advice to special risk groups, without means
test.[47] Provincial funding of this service has diminished
relatively since the National Health Insurance programs
were introduced because it is not matched by Federal
contributions. By 1955, however, the main infectious
diseases had been brought under control and this left
more for other activities. It took the "New Perspective"
report to give the next push to general preventive
measures and these are not yet well developed despite
efforts at restructuring DNHW and provincial Ministries
to encourage shifts in direction.

The last five years have been characterized by gap
filling for elderly and handicapped. This
reorientation forced more attention to be paid to all
health promotion programs: fitness campaigns, seat belt
legislation, stress reduction; and environmental
issues are now more popular (though the last less so since
unemployment has threatened). Experiments in mass
screening have been abandoned in favour of screening
on indication but there is usually no problem for patients
in getting to primary care to check symptoms as early
as they wish to do so (except where user charges have
crept back in). Health education takes place in
doctors' offices, hospitals and public health settings
and there are self-help and crisis intervention groups too,
but much more invention is needed in this area.
Coordination occurs less by grouping diseases than by
extending programs for population groups. Special programs
for the mentally ill, drug addicts, etc. which were
started separately have been, where possible (i.e. for
acute care stages) integrated into the general system of
care.[48] Attention is also being given to 'normalization'
of chronically ill or disabled persons.[49]
On the other hand, special programs may be put on for
identified risk groups such as immigrants or native
peoples.[50]

4.2. Primary Health Care

Since the government funding of Medicare was
introduced, primary health care has been readily available,
except (1) where the distribution of physicians has been
unsatisfactory, (2) where user charges have been allowed
to be made.[51] The system of admitting to hospital privileges
permits PHC physicians to follow patients through
hospitalization periods and consultants are expected
to keep in close contact. Referrals tend to be to personally
known specialists, not necessarily to the nearest regional
consultants. Whilst this is food for inter-personal
relations it is wasteful organizationally. Support of PHC
is better at referred levels than in provision of social
community support programs which are only just beginning
to be thought through.

4.3. Coherency

National health insurance programs were based on
concepts of equality of condition for consumers.
Provincial governments had to agree to the principles of
portability, reasonable access, universal coverage,
comprehensive coverage, public administration. They were
also based on concepts of equality of opportunity for
providers - reasonable compensation, uniform terms and
conditions. Many consumers thought they had been promised
equity or "fair shares". To satisfy these somewhat
conflicting concepts is difficult in a totally
controlled system. To do so in a "subsidized
entrepreneurial" system even more so.

5.0. Health Services Systems Management

The Canadian health services system has failures in
organizational design as outlined above. This makes it
more difficult to manage than a well designed system, yet
management is still relatively undeveloped.

5.1. Leadership and Management Skills

Politicians are attempting to put their own houses
in better order under the pressures of the present
recession. They find it difficult (as do hospital
trustees) to draw clear distinctions between policy
making and management.

The transition of the government bureaucracies from
clerical machines to professional advisory groups is far

from complete (see 1.2) and the politicians' involvement
is not enabling them to complete the transition in many
provinces. One of the reasons is the absence of
satisfactory mechanisms for linking political and
bureaucratic decision-making.

There are difficulties too in deciding how to use
clinician bureaucrats, whether as line managers or
staff advisors. The training of health service
administrators has been improving in the last ten years[52],
but there are many who are floundering in their jobs.
Neither they nor their employers can grasp all the
issues, set broad objectives and measurable goals without
difficulty. And if they are good at objective setting and
monitoring then they may have problems in
negotiating with medical staff or hospital employees.

Capacity for innovation is certainly there but
corporate planning has hardly begun to develop and program
evaluation is relatively new.

In the present recession there is much discussion
of tighter control but it is clear that managers at all
levels must negotiate objectives and their implementation,
rather than develop control mechanisms and try to impose
them from above.

5.2. Decision-Making Processes

1. Saskatchewan has provided a model for rational health
   planning since 1944[15] but federal funding offers have
   often cut across rational approaches. The Saskatchewan
   data collection system has formed a model for other
   provinces (recognized later to be inappropriate
   to their different demographic needs as in the
   development of a B.C. Hospital Insurance system). The
   thoroughness of this information system has been seen
   as a threat to professional autonomy,[8, 10] especially
   as it has enabled some bureaucrats in the Medical Care
   Commission to pinpoint abuses in elective care (e.g.
   excessive surgery). Despite the thoroughness of information
   it has been found impossible to close hospitals readily.

   Health policy formulation in Canada is, as discussed above,
   a matter of trade offs and bargains. Initiatives are
   taken by different interest groups at different times,
   legitimacy established, feasibilities worked out
   and supports built up. Programming is relatively
   sophisticated, budgeting very unsophisticated
   (see 3.1).

2. There are basic conflicts between managers and
   clinicians on issues of implementation and
   realization. Rationing has barely been discussed and
   certainly not in clinicians' terms.[53]

3. Processes of monitoring and evaluation necessarily
   follow processes of objective setting and
   establishing proper inputs. These prior processes are
   still often unclear.

   Monitoring of capital budgeting is effective but often
   implemented by petty bureaucrats; similarly it goes with
   operational activity. Trees are monitored rather
   than woods evaluated.

   Questions about maximal, optimal, minimal standards
   are not yet being asked though new questions about
   value for money are now emerging.

4. Appropriate information flows depend upon appropriate
   questions. Since questions are presently being reformulated,
   appropriate information is not always available.
   Governments are concerned about improving 'economic
   forecasting' and other research capacity.

5.3. Mechanisms of Regulation

A number of Acts have been revised in the last twenty
years, but there is need to improve the writing of
legislation which in some Acts is quite primitive.

Particular attention has been paid in recent years
to legislation regulating the professions,[30] and many
provinces have revised or intend to revise this[54] but
much depends on the concept of professional rights and
responsibilities in present day society. These are not
clear.

Equally, standards and norms of practice are uncertain.
These are usually left to peer reviews. There are
pressures, however, to change the legislation relating
to control over ethical questions (e.g. defining death,
in vitro fertilization) and to tighten up on resource
allocation to physicians whose volumes of work are
adjusted to meet competition.

Postscript

In April 1982 the Constitution Act was passed. This
followed a challenge to federal government by several pro-

vincial governments in the Courts. The Supreme Court Justices advised that, while Parliament had the power to legislate the Bill into law, the government would be illadvised to put it through without negotiating changes with the provinces. Thus federal-provincial conferences have been given a new authority.

In May 1982 the federal Minister of Health announced that she was not going to withhold payments to the provincial governments which permitted extra-billing (since patients rather than providers were more liable to suffer). Instead, she intended to redraft legislation to tighten up the definitions of the principles enunciated in the Medical Care Act 1966, viz. universality, accessibilty, comprehensiveness and portability with a view to improving more equitable distribution of health resources. This move is one more indication of the power shift from federal to provincial governments in the last decade.

It should be noted that it is very difficult to obtain information about health policy developments in Canada since much of the information is prepared for private circulation within governments or is available only on request. Unless one is able to travel between provinces it is not easy to keep up with what is going on. In consequence this paper has errors and ommissions but, it is hoped, deals with general trends. A series of papers on policy issues for the government of British Columbia [55], prepared by a study group at the University of B.C. is to be published in the fall of 1982 with the intention of promoting discussion across Canada about similar problems in other provinces.

References

1. Canada. Year Book 1980-1981. Ottawa: Information Canada 1981.
2. Alford, Robert R. Health Care Politics, Chicago : University of Chicago Press, 1975
3. Titmuss, Richard M. Social Policy, London: George Allen & Unwin, 1974.
4. Marchak, M. Patricia. Ideological Perspectives on Canada, Toronto: McGraw Hill Ryerson, 1975.
5. Lowi, Theodore J. "American Business Public Policy, Case-Study and Political Theory", World Politics, Vol. 16, 1964, pp. 677-715.
6. Hirsch, Fred. Social Limits to Growth, Thetford , Norfolk, England: Lowe & Brydone Printers Ltd. 1978.
7. McKinlay, John B. "The Business of Good Doctoring or Doctoring as Good Business: Reflections on Freidson's Views of the Medical Game", International Journal of Health Services 7:3, pp. 459-83, 1977
8. Johnson, Terence J.,Professions and Power, London: Macmillan 1972.

9. Lang, Ronald W. The Politics of Drugs, Lexington, Mass.: Saxon House, 1974.

10. Badgley, R.F. and S. Wolfe.,Doctor's Strike, Toronto: Macmillan, 1967.

11. Canada (Hall), Royal Commission on Health Services, Ottawa: Queen's Printer, 1964.

12a. Andreopoulos, S., ed. National Health Insurance: Can We Learn From Canada?,New York: Wiley, 1975.

12b. Anderson,Donald O., "Canada: Epidemiology in the Planning Process in British Columbia: Description of an Experience with a New Model", in White, Kerr L. and Henderson, Maureen M. (eds.) Epidemiology as a Fundamental Science, New York: Oxford, 1976, pp. 44-59

13. Canada(LaLonde), Department of National Health and Welfare, A New Perspective on the Health of Canadians. Ottawa: Information Canada, 1974.

14. Labonté, R. and Penfield P. Susan, Health Promotion Philosophy : From Victim Blaming to Social Responsibility: A Working Paper, Vancouver: B.C. Government Health Branch 1982.

15. "The Sigerist Report" in (ed.) M.I. Roemer, Sigerist on the Sociology of Medicine, New York: M.D. Publications, 1960

16. Gil, David G.,A Systematic Approach to Social Policy Analysis, Waltham, Mass.: Brandeis University, Working Paper No. 1, Social Policy Study Program, 1970, pp. 1-26

17. Campbell, A.G. et al.,"Changing Strategies for British Columbia Health Management: From Program Development to Cost Control" Vancouver Health Services Planning Discussion University of B.C., May 1981.

18. Canada, D.N.H.W. Report of the Task Force on the Cost of Health Services, Ottawa: Information Canada, 1970.

19. Newman, Peter C.,The Distemper of our Times.,Toronto: McClelland and Stewart, 1966.

20. Carter, Novia.,Trends in Voluntary Support for Non-Governmental Social Service Agencies, Ottawa: Canadian Council on Social Development, 1974.

21. Joint Commission on Accreditation of Hospitals, Accreditation Manual for Hospitals, J.C.A.H. Chicago: 1979.

22. Macdonald, Alastair, Medical Malpractice Litigation in Canada, Vancouver: unpublished M.Sc. thesis, University of B.C., 1977.

23a. Canada: Dept. of National Health & Welfare, National Conference on Assistance to the Physician, Ottawa: DNH & W. April 1971.

23b. Canada: Dept. of National Health & Welfare, Report of the Committee on Nurse Practitioners, Ottawa: Information Canada, 1973.

23c. Sackett, D.L., Spitzer, W.O., Gent, M., et al.,"The Burlington Randomized Trial of the Nurse Practitioner: Health Outcomes of Patients", Annals of International Medicine, February 1974. pp. 137-142.

23d. Canada (Hastings): Department of National Health & Welfare, The Community Health Center in Canada, Ottawa: Information Canada, 1972-73 (3 vols.)

24. Evans, R.G. "Beyond the Medical Marketplace", Chapter 3 in ed. Andreopoulos, Spyros, National Health Insurance: Can We Learn From Canada? Toronto : Wiley, 1975.

25. Canada : Dept of National Health and Welfare, Annual Reports, Ottawa: Information Canada.

26. Quebec (Castonguay-Nepveu) Report of the Commission of Inquiry on Health and Social Welfare, Quebec: Government of Quebec, 1966-70.

27. Parsons, Talcott, The Social System, London: Tavistock Publications Ltd., 1952

28. In September 1981 there were 6690 "directory active" physicians in B.C. The population of B.C. was calculated at 2.667.200. University of B.C., Office of the Coordinator of Health Sciences: Health Manpower Research Unit, Roll Call, Vancouver: HMRU, 1981.

29. Miles, Robert H., Macro Organizational Behaviour, Santa Monica: Goodyear Publishing Co., 1980.

30. Canada Economic Council of Canada, Reforming Regulation, Ottawa: Information Canada, 1981.

31. Canada, Federal and Provincial Manpower Working Groups are actively reviewing prospects for improving planning activities.

32. Personal communication for Jack Bainbridge, Assistant Deputy Minister, B.C., who set up the Long Term Care Service in 1977.

33. British Columbia, Hospital Role Study, Phase 1, Victoria: Ministry of Health, August 1979.

34. Canada: Science Council of Canada, Science for Health Services, Ottawa: Information Canada. 1974.

35. Detwiller, L.F., "The University of British Columbia Health Sciences Centre - A Coordinated and Integrated Training Program for Members of the Health Team of Today." Journal of the Royal Society for the Promotion of Health. Vol. 83, No. 6. December 1963. p. 293 seq.

36. Allen, Elizabeth., Case Study in (eds.) A. Crichton and S. King, Management of Health Service Organizations, Ottawa: Canadian Hospital Association. In press.

37a. Badgley, R. and Smith, D.A., User Charges for Health Services, Toronto, Ontario: Economic Council, 1979.

37b. Barer, M.L., Evans, R.G. and Stoddart, G. Controlling Health Care Costs by Direct Charges to Patients. Snare or Delusion?, Toronto, Ontario: Economic Council 1979.

38. Canada, Canada's National Provincial Health Programs for the 1980's, Saskatoon: Health Services Review '79, 1980.

39. Canada, Parliamentary Task Force: House of Commons, Fiscal Federalism in Canada, Ottawa: Information Canada, 1981.

40. Hardwick, Jill, M., The RNI as Case Mix Indicator: Promise

and Performance, Vancouver Unpublished, M.Sc. thesis. University of British Columbia, 1982.

41 The B.C. Government has just introduced changes into its Long Term Care Policy to meet financial stringencies, 1981

42. Taylor, Malcolm G.,Health Insurance and Canadian Public Policy, Montreal: McGill, Queens, 1978.

43. In Nova Scotia in pre-World War II years the miners hired their own contract physicians.

44. A review by Govan, Elizabeth, S.L. Voluntary Health Organizations in Canada, Royal Commission on Health Services, Ottawa: Queen's Printer, 1966.

45. Crichton, Anne.,Community Health Care, Unpublished commissioned report prepared for Health Services Review '79, 1980

46. Crichton, Anne et al.,Health Seen in N.W. B.C.,Vancouver: Department of Health Care and Epidemiology, mimeo, 1975. (limited circulation).

47. Hastings, J.F.L. and Moseley,W.,Organized Community Health Services: Ottawa, Queen's Printer, 1966.

48. Canadian Mental Health Association, More for the Mind, Toronto: CMHA 1963.

49. Wolfensberger, Wolf, The Principle of Normalization in Human Services, Toronto: National Institute on Mental Retardation, 1973

50. The Vancouver Health Dept. (to take one example) has developed a series of special risk programs for early diagnosis of diseases of the elderly, to prevent low birthweight, etc.

51. Beck, R.G.,"Economic Class and Access to Physicians Services under Public Medical Care Insurance", International Journal of Health Services, Vol.3, No. 3. 1973. pp. 341-355

52. W.K. Kellogg Foundation, Unmet Needs. Education for Health Services Administration in Canada, Ottawa: Canadian College of Health Services Executives, 1978.

53a. Parker, R.A. "Social Administration and Scarcity", Social Work, April 1967. pp.9-14

53b. Crichton, Anne, Health Policy: The Fundamental Issues, Ann Arbor: Health Administration Press, 1981.

54a. Quebec. The Professions and Society. Report of the Commission of Inquiry into Health and Social Welfare, Vol. VII, Tome 1, Part Five. Government of Quebec, 1970.

54b. Ontario, Report of the Committee on the Healing Arts, Queen's Printer, 1970.

55. University of British Columbia Health Policy Study Group, "Current Issues in Health Policy for the Government of British Columbia" Health Management Forum: Fall 1982 (in press).

THE HEALTH SERVICES SYSTEM OF THE UNITED KINGDOM

Mr. Keith Barnard
University of Leeds

Dr. David Pendreigh
University of Edinburgh

United Kingdom

Introduction

The United Kingdom is composed of four countries:
England, Wales, Scotland and Northern Ireland. For many
governmental functions the four are in fact one country.
But there are certain exceptions and for some areas of
public administration there are separate structures for
all four countries. Health is one such area and this paper
focusses on England, the largest in terms of geography
and population (46m.) and subsequently on Scotland, the
second largest.

However, despite administrative differences, often
of detail, between the four countries, the values, goals
and objectives underlying the health service system are
the same. These were laid down in the original statute,
the National Health Service Act 1946*. Although the Act
and subsequent related legislation were consolidated in the
NHS Act 1977, the original intentions have remained unchanged.
The Act requires the Minister i.e. the Secretary of State
for Social Services to promote a comprehensive health service
designed to secure improvement

a. in the physical and mental health of the people;
b. in the prevention, diagnosis and treatment of illness.

Section 3 of the 1977 Act states "It is the Secretary
of State's duty to provide ..... to such extent as he considers

---

* There was a parallel NHS (Scotland) Act 1947

necessary to meet all reasonable requirements .....,
facilities for the care of expectant and nursing mothers
and young children as he considers are appropriate as part
of the health service ..... such facilities for the prevention
of illness, the care of persons suffering from illness and
the aftercare of persons who have suffered from illnesses
as he considers are appropriate as part of the health service".

The Royal Commission on the NHS set up in 1976 to review
the functioning of the Service, reporting in 1979,
summarised the objectives of the NHS as they saw them as:

- to influence individuals to remain healthy;
- to provide equality of entitlement to health services;
- to provide a broad range of services of a high standard;
- to provide equality of access to these services;
- to provide service free at the time of use;
- to satisfy the reasonable expectations of its users;
- to remain a national service responsive to local needs.

While both the legislation and the Royal Commission's
perceptions are capable of more than one interpretation
operationally, the underlying philosophy of the NHS is clear
enough and it is one which has been shared by political
parties in government and opposition across the political
spectrum.

Thus the health service system reflects the general basic
political configuration. The UK is a multi-party democracy
with identified organised political parties which together
span the whole political spectrum but only the moderate
left centre and moderate right parties are normally
represented in parliament. Over the past twenty years or
so there have been frequent changes of ruling party which
has created some discontinuity in public policy: these have
been mainly on the management of the economy and the nature
of government control and support of industry. Ideological
differences between parties in health matters have surfaced
from time to time but have not so far been critical to the
long term development of the health service. There is some
evidence, not conclusive, that this underlying consensus on health
policy may now be disturbed.

The UK has a mixed economy with a large public sector
covering some major manufacturing and trading enterprises and
several public utilities as well as a wide range of publicly
provided social services including medical care. The general
character of the public/private mix of the economy emerged after
the Second World War with some periodic expansions
of the public sector occurring subsequently under various

governments. The present UK government came to power in 1979 committed to contain and indeed to contract the public sector by various measures including the withdrawal of the monopoly enjoyed by certain public enterprises.

Consistent with this ideological position, the Government has in some of its legislation and in its public utterances given its support for a major expansion of private health services. Although private health services are in some ways seen currently as a major growth service industry, the growth has to be seen in the context of its original small base and a change of government could bring an end to official support. In short, the private sector is an issue of ideological dispute which could remain on the public agenda for some time.

Health Services System Structure

Notwithstanding the emergence of a private sector (covering broadly those coming to the country for medical treatment, UK citizens insured for private treatment, usually surgery, and nursing care for the elderly in private nursing homes), it is the public sector organisation which dominates the system.

As indicated above, the legislation obliges the Minister to develop a system for the whole population (and on any reasonable assumptions or calculations well over 90% make use of the NHS). In fulfilling this obligation, his principal tasks are to secure funds from the Treasury as approved by parliament as one of the government's spending programmes; to allocate the funds to the NHS to run the services it provides; to issue general policy guidance which he expects the NHS to interpret and implement in each geographical area according to local circumstances; and to monitor the NHS to ensure that local services are functioning effectively and efficiently in line with his policies.

In so far as the promotion and regulation of the private sector (e.g. approval of certain private hospital capital developments) are also within the Minister's sphere of responsibility, there is clearly one focal point for national policy for what is conventionally regarded as health care. When the position of Secretary of State for Social Services was first created in 1968, the intention was that he would be the Minister responsible for health and for social security. Two Ministries with those titles were combined as the Department of Health and Social Security which he was

to preside over. His title 'Social Services' was to indicate
that besides his direct responsibilities he was also being
asked to take on a general co-ordinating role embracing
other social services such as education which are the
departmental responsibilities of other Ministers.

It is informally understood that this co-ordinating
role has not subsequently been actively discharged by
successive Secretaries of State and it may well be that the
role was contested by other Ministries and their Ministers.
There is some evidence to support that assumption. In 1970
a Central Policy Review Staff (CRPS) was created, in effect,
to assist the Prime Minister in matters of policy formation
and related administrative matters. One issue which exercised
the CRPS in the early 70's was the need for a joint approach
to social policy (JASP) whereby social problems for which
a public policy response was appropriate, would often span
the boundaries of government departments and therefore an
interdepartmental policy would be needed requiring collaboration
between departments in its implementation and having
organisational and budgetary consequences. It is a reasonable
inference that if the Secretary for Social Services had been or
been able to exercise his co-ordinating role, CRPS would not
have deemed it important to press the joint approach on
Ministers. By 1975, they had engaged the interest of Ministers
sufficiently for the publication of their proposals to be
sanctioned by the government. They appeared in a monograph
'A Joint Framework for Social Policy' having previously
been extensively leaked to the press. At the time they
promised an imaginative breakthrough in public administration
and policy formulation and drew comparison with the already
published Canadian document 'A New Perspective on the
Health of Canadians' which focussed more narrowly on health
policy but which was founded on essentially the same basic
assumptions of the nature of social problems.

Unfortunately, the ideas of the joint approach did not
survive the test of exposure in the area of government
administration, probably because it directly addressed
the de facto autonomy of individual departments and was not
an issue of sufficient overt political importance to attract
the continuing interest of Ministers. Indeed the whole
issue of the machinery of government which attracted
considerable ministerial interest and intellectual investment
by others in the early years of the 70's was no longer a
matter of concern by the end of the decade; except in so far
as there was a reaction against the 'growth of government'
accompanied by a search for savings in public expenditure.
With the failure of JASP to command serious attention, there
were no further public attempts to co-ordinate social policy

in the same overarching fashion. Rather, it must be assumed, efforts of co-ordination have proceeded, presumably largely in a piecemeal fashion, through the normal unpublicised processes of inter-departmental liaison and Cabinet and Cabinet Committee meetings involving both Ministers and civil servants.

Department of Health and Social Security (DHSS) and the National Service

The distribution of functions between Ministries gives the Secretary of State for Social Services three major responsibilities:

(I)   Social Security, i.e. the administration of various income maintenance schemes laid down by statute including old age pensions and cash benefits to ameliorate loss of income in certain circumstances such as sickness and injury. These schemes are administered through a nationwide network of local offices of DHSS which is managed in a simple hierarchical fashion in the name of the Secretary of State.

(II) Personal Social Services. The Secretary of State does not provide directly any on the services so labelled (help for various categories of people in need or distress which may take the form of domicillary support or institutional care) but gives policy guidance and exercises a range of controls over quasi-autonomous Local Authorities. These are political elected bodies which provide a range of public services including social services. They have tax raising powers called 'rates' (property taxes) but receive the larger part of their operating budget through central government grants.

(III) The National Health Service covering medical services. These are provided formally in the name of the Secretary of State but in a fashion which may be described as a hybrid of (I) and (II). The Secretary of State appoints corporate bodies, Regional Health Authorities, which are in law both his agents and also quasi-autonomous. RHAs in turn appoint District Health Authorities for each local area which are similarly enjoying an ambiguous relationship with the higher tier in the structure, in their case, the RHA.

Recent government initiatives, however, have suggested
that present Ministers are keen to emphasise the
accountability of subordinate to superior tiers
in the system rather than the element of autonomy.

The functional limits to the NHS were first laid down
in a public government discussion document in 1969 and the
position adopted then was subsequently adopted in legislation.
The NHS would include those services where the skill of the
personnel was medical or medical related (e.g. nursing).
All staff with other skills (e.g. social work) would be
employed by other agencies and their work would be the
responsibility of that agency even though their normal place
of work would be a health institution. At the same time other
agencies could no longer directly employ health workers
but the local Health Authority was charged with providing
medical advice whenever required as by the various local
authority departments including Social Services and
Environmental Health.

Finally, reference should be made to the existence of
special services and facilities outside the NHS. These
include special hospitals administered directly by DHSS
for mentally disordered offenders, the Prison Medical
Service, and the medical services of the Armed Forces
which are also available to the dependents of servicemen and
to the general public, in effect, as an additional resource to
the NHS. Occupational Health does not come within the
NHS and any formal provision is at the discretion of the
individual employer. There is, however, a statutory Health
and Safety Executive under the Department of Employment
charged with monitoring and enforcing measures to control
hazards in the workplace; and also an official Employment
Medical Advisory Service for the benefit of employers and
their workforce.

The rest of this paper will refer only to the National
Health Service.

NHS Administration

Reference has already been made to the functions of the
Secretary of State and the DHSS and their relationships with
NHS authorities and local authorities. In effect, DHSS
is run as two departments which only become one at the highest
level, i.e. Social Security and Health and Personal Social
Services which are treated together in terms of developing
government policy and identifying national priorities.
This is in recognition of the close relationships between

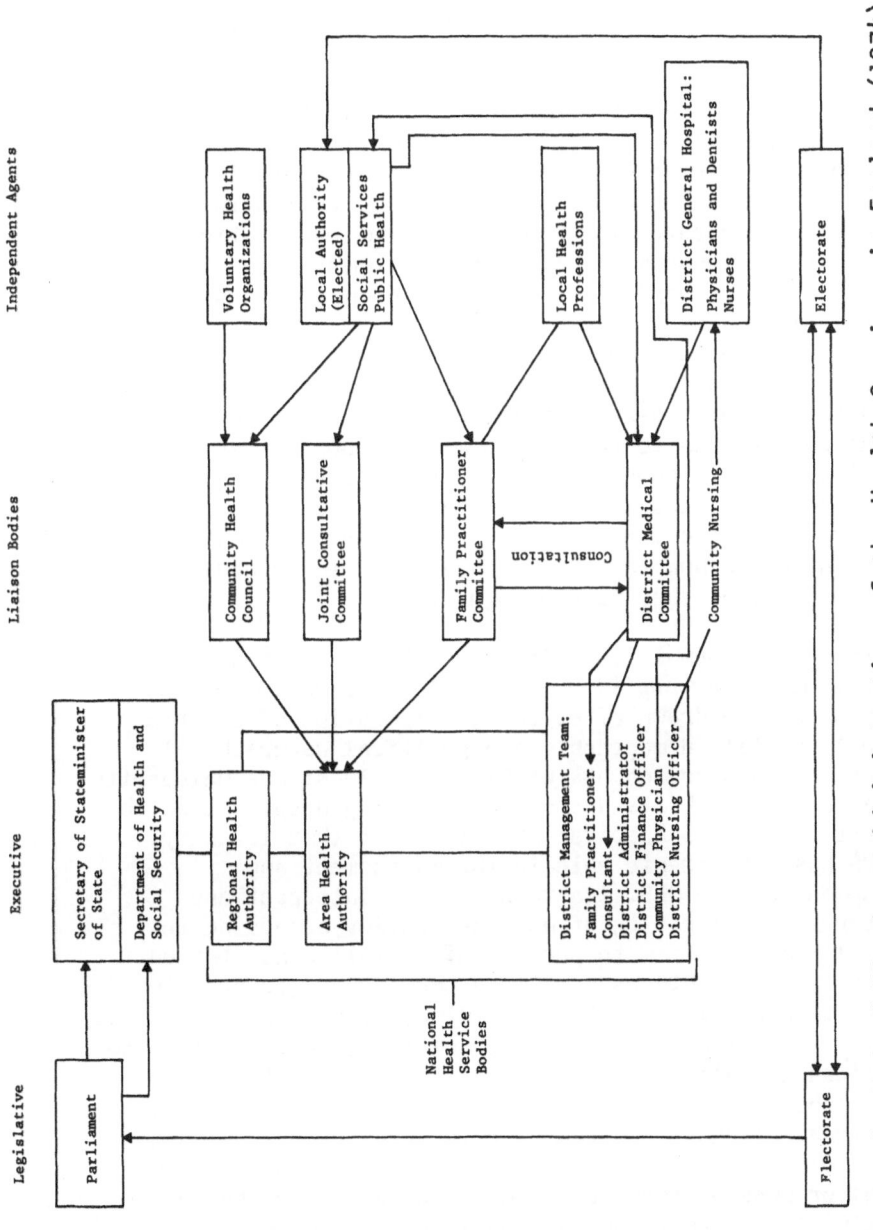

Figure 1 United Kingdom: Administration of the Health Services in England (1974).

the social and health needs of many often highly dependent groups
in society such as the elderly and the handicapped.
Government thinking on such issues is encapsulated in
policy guidance circulars and other documents which are
issued from time to time to health and local authorities,
but as indicated earlier, the DHSS only enjoys a direct
corporate relationship with the NHS.

In essence then, the role of DHSS is to formulate advice
on and subsequently implement Government policy.
Although the Secretary of State directly accountable to
parliament, has the necessary powers of direction in reserve,
DHSS normally operates through persuasion, exhortation and
to some extent inducement. Its key instruments are its
approval of major capital developments proposed by the Service,
its control (in association with the profession) of the
expansion of medical manpower, and its allocation of
virtually all financial resources to the Service. Against
the background of those controls, DHSS requests the NHS
'field' authorities to review and make proposals for the
provision and development of services in line with government
policy and priorities and within the financial assumptions
they have been given. Subsequently Ministers and their
officials review how far the NHS has moved in accordance
with the guidance they have been given.

The DHSS' normal link with the NHS is with the fourteen
Regional Health Authorities (RHAs). The boundaries of each
region have been drawn after due consideration of a variety of
physical, social, demographic and political geographical
factors and each region has within it at least one University
medical school with its associated teaching hospitals which
are part of the NHS hospital provision for that locality.
The RHS has a strategic role in the management and
development of the NHS. Its tasks are to interpret and
consider the application of national policy according to
regional circumstances to prepare after full consultation
with local level interests, a long term strategic plan for the
region; to plan directly the capital developments and medical
manpower expansion which arise out of the strategic plan; to
make financial allocations to District Health Authorities
to run local services and to require DHAs to prepare local
operational plans in accordance with the strategic plan and
realistic financial assumptions. Lastly RHAs are expected
to monitor DHAs performance. Except for a limited range
of activities where technical considerations or expected
economies of scale so determine, RHAs are not involved in
the management of operational services which are reserved
to DHAs. Their control over major capital developments and
the location of specialist medical manpower are key factors

determining the shape and scale of hospital and other services that a particular DHA has to manage.

The structure of the NHS set up by the NHS Act 1946 had separate administrations below the Ministry for hospitals, a range of personal health and environmental services (provided by the local government authorities), and the general medical and dental services provided by private practitioners referred to as independent contractors. This tripartite structure was arrived at, partly under the influence of previous history, partly by what was acceptable to the medical profession, and partly by the organisational necessities of launching the NHS. It was at the time seen as an effective political compromise but not ideal, in that it carried the risks of fragmented separate development of parts of the service in any locality, and depended almost entirely on the motivation of individuals to promote the effective coordination of services at the local level. By the 1960s such potential or even actual disadvantages were seen as important enough to warrant major reorganisation of the NHS.

In 1974 after many years of public debate the NHS was reorganised into Regional Health Authorities (successors to the Regional Hospital Boards established under the 1946 Act) and Area Health Authorities (whose boundaries were determined by those of the matching elected Local Authority responsible for education and social service provision). These two statutory tiers of Region and Area absorbed the functions of the previous hospital and local authority health services. The administrative functions relating to the services of the independent contractors were kept separate and given to a body, the Family Practitioner Committee (FPC), which received its budget directly from DHSS. The FPC covered the same geographical area as the AHA, but although there were some built-in organisational links between FPC and AHA, the FPC was able to act almost as an autonomous body.

Thus the underlying purpose of the 1974 NHS Reorganisation to create a unitary health authority at local level was thwarted at the outset. Additionally, as already mentioned, political and professional considerations had determined that health and personal social services should be administered by separate agencies notwithstanding the frequent interdependence of the two in the interests of patients and clients. This separation therefore necessitated the establishment of statutory machinery, a Joint Consultative Committee, at local level to foster collaboration between the two agencies in terms of joint planning and the

coordination and integration of operational services.
Subsequently more detailed arrangements were developed
to bring together the managers and operational staffs of
the two agencies functioning as a joint care planning team
(JCPT) under the aegis of the J.C.C. The JCC was necessarily
merely a liaison body as decicions on resource allocation
and investment would be taken by the two parent agencies.

Two other features of the 1974 Reorganisation should be
mentioned. These were, firstly, the establishment of a
variety of advisory and consultative committees to give
different interest groups a formal opportunity to exercise
some influence on the decision making process. This machinery
was established for the medical profession, other groups
of health workers, and at local level representation
(nominated rather than elected) of the general public
(known as Community Health Councils). The second feature
was the necessity to establish an additional tier in the
structure i.e. the Health District as a sub-unit of the Area.
About two thirds of all Areas were sub-divided into two
or more Health Districts for management purposes as the Areas
and the services provided within them were regarded as
too large to be managed as one unit. A Health District
was determined by the population (say 200.000 but it could
be larger or smaller) served by a given set of hospital
and community services providing primary care and the basic
specialities of secondary care. These services were run
by a multi-disciplinary District Management Team (DMT)
accountable to the AHA.

While all the features of the 1974 Reorganisation
appeared initially to have a defensible logic and surface
'reasonableness', they quickly became the subject of fierce
criticism. There were various contributory factors. The change
over from the old to the new authorities was particularly
disturbing for many senior officials who were expected to
learn instantly new roles and ways of working. The creation
of the new administrative superstructure appeared to
quickly denude the operational levels of the service of all its
experienced managers. The change destroyed the well established
network of influence and information to which medical
staff had become accustomed and knew how to exploit. The
destruction of that network, the complexity of the new
arrangements and the apparent diffusion of responsibility
for decision making left senior medical staff extremely
frustated and fierce opponents of the structure. Before
long they were making that opposition very public.

The 1974 structure had been advocated as the basis for
future rational development and management whereby decisions

on priorities and resource allocation could be based on
a well informed assessment of 'needs'. The existence of
consultative and collaboration machinery with interested
parties and other agencies would enable agreement to be
reached on the right policies to be followed. In the event
the underestimate of the perpetuation of sectional interest
(intensified by the slowing down in the growth of resources
for the NHS) and the sheer complexity of the management
arrangements demonstrated the apparent naievity of their
proponents.

It will remain a matter of speculation whether, given
time, the 1974 arrangements could have been made to work
or amended to become more effective by an evolutionary
process. The Government elected in 1979 was already
committed to introduce legislation to simplify the
structure. Their changes became effective in April 1982.

NHS Restructuring 1982

The 1982 reforms had as the Government's declared
objectives to simplify the structure and to delegate and
improve the quality of decision making. Their proposals
removed the Area tier and retained Regions and RHAs and
Districts which were to be governed by District Health
Authorities (DHAs), in effect the successor bodies to AHAs
having essentially similar functions and accountable to
RHAs for the management of operational services. 90 Areas
were replaced by 193 Districts. The apparent intention was
also to reassert the dominance of the hospital in the
interests of efficient institutional management rather
than an appropriate balance of services planned according
to perceived 'needs' and organised to provide integrated
care. Such suspicions were aroused by the phrase used in an
early government document 'hospitals and their associated
community services' implying that the hospital was the
central feature of the NHS. In terms of financial and
manpower resources consumed this is unquestionably true,
but the phrase was taken to imply that good husbandry had
replaced 'service developments' as the driving force of
NHS management proclaimed in 1974. This interpretation was
subsequently denied by ministers who suggested that the
offending phrase was little more than a drafting point.

Nevertheless the circumstantial evidence supporting the
interpretation was strong, since the decision to abolish
the Area tier broke the link between a health authority and one

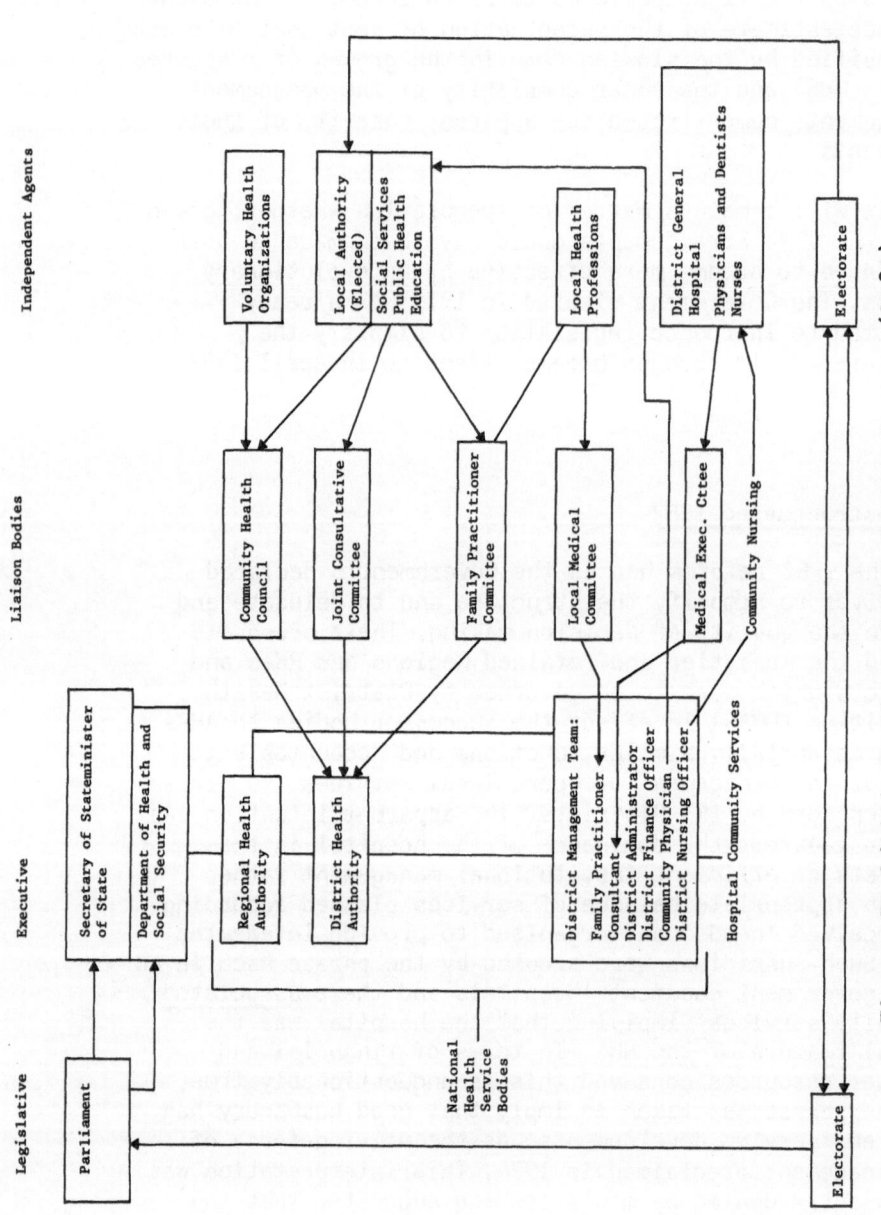

Figure 2 Administration of the Health System (1982).

matching elected local authority. Henceforth local authorities
would have to develop collaboration arrangements with a number
of DHAs which, it may be anticipated, will exacerbate problems
of agreeing on decisions of priorities, plans, and resource
allocation. Further circumstantial evidence of Ministers' lack
of commitment to integrated comprehensive care was provided
by their decision to introduce legislation when practicable
to completely separate constitutionally Family Practitioner
Committees from Health Authorities. At the time of writing
it is still too soon to see whether such fears and accusations
are well placed or not. But the decision to introduce a
second set of major organisational changes after eight years
compared to the twenty six years which elapsed between
1948 and 1974 should prompt questions in the minds of students
of systems as to: what pace and degree of change such a complex
organisation can be subjected to; what criteria should be
applied when considering and implementing change; and the risks
to the underlying stability of the system when politically
determined decisions critical to the system override managerial
considerations such as the need to minimise turbulence for
key personnel.

Health Resources

Manpower

   Despite its monopsonic position as virtually the sole
employer, the NHS is commonly observed to have failed to match
the demand for and supply of trained health manpower, both
in respect of the geographical distribution of trained
workers and the number of trained workers. This is in spite
of a notable increase in numbers since the start of the
NHS in 1948.
Medical manpower, perhaps, not unexpectedly has attracted the
most attention and debate. The medical profession engaged
in clinical practice can be divided into two main groups,
general medical practitioners (GPs) working in either their
own or NHS premises such as health centres, and the specialties.
GPs account for over 1/3rd of all medical practitioners.
Until recently formal training was voluntary but was gradually
becoming more organised and systematic; now all entrants into
general practice must complete a prescribed form of
vocational training. In contrast the specialties based in
hospitals have long been characterised by a highly structured
sequence of apprenticeship culminating in a tenured
position as a Consultant Specialist.

   For almost the first 20 years of the NHS, hospital
practice in a specialty was considered to be the aim of any
ambitious young doctor. Entry in GP was commonly seen as

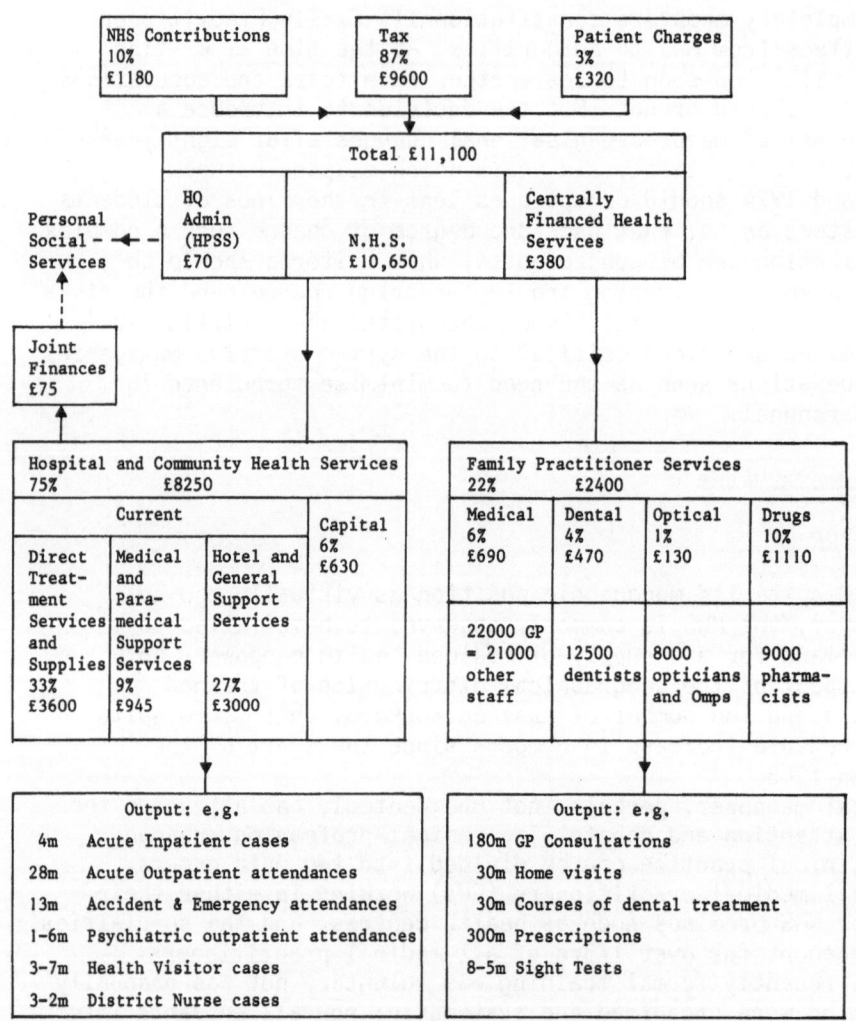

Figure 3   N.H.S. Overview 1981-1982.

failure to succeed in a specialty. The relative
financial rewards also reflected that value judgement.
Attitudes changed from the mid 1960s onwards and there has
been, for some years now, a healthy number of medical
graduates who choose General Practice and since a major
reshaping of the renumeration package for GPs in 1967
(a basic allowance, capitation fees and some item of services
payments) earnings of GPs significantly improved.

Control of numbers of GPs in the NHS is the responsibility
of a central Medical Practices Committee. However, the
control exercised is distinctly modest: specific localities
are in effect designated as under-or over-doctored.
Over-doctored areas are barred to new GPs wishing to practice,
while financial inducements are offered to bring them into
under-doctored areas. There is no evidence that this tactic
has had a major effect of ensuring full equality of access
to primary medical care although it is a reasonable
assumption that it may have promoted greater equality than
might otherwise have prevailed.

The growth of the hospital based specialties is controlled
by a central, and in each region a regional manpower committee.
The intention is to ensure a desired rate of growth
geographically and by specialties. This also implies a balance
between the tenured consultant positions and the junior,
fixed term training positions. It is here perhaps that the
failure in manpower planning (usually focussed as the
problem of so called shortage specialties and shortage regions)
has so far seemed intractable. There is an obvious conflict of
interest between the Consultants who want to limit the number
of tenured positions and expand the junior grades to give
them assistance, and the junior doctors who want an expansion
of the consultant posts to create more career opportunities
for them. Current government policy is to expand the number
of consultant posts but it remains to be seen whether this
policy can be implemented to the extent required. As has been
observed, every study of manpower over the past 20 years had
recommended the expansion of the Consultant grade, while
the unpopular specialties of the past remain unpopular and
no acceptable inducements have been discovered.

Meanwhile, there is growing agitation among junior
doctors over their career prospects and even the risks of
medical unemployment. This revives a debate of the 1950s
when the age structure of the profession and career
prospects led to a successful agitation for a reduction in
the intake to medical schools. It was an early mark of the
uncertainty of medical manpower planning that the evidence,
including that of emigration, within 10 years was pointing

in the direction of a significant expansion of medical school
intake, which policy has remained to the present time.
The Royal Commission on the NHS in its Report (1979) concluded
that the planned output of medical schools was about right.

In contrast to varying and perhaps sophisticated
thinking about medical manpower, the view of nursing manpower
was for many years unchanging and simple: there was a
shortage. This was despite steady growth in numbers throughout
the period of the NHS although the growth also needs
to be seen in a context of improved conditions of service
for nurses including reductions in the number of hours
worked. The common wisdom changed in the late 1970s when
a surplus of trained nurses came to be feared. This changed
again as authorities reported difficulties in nurse
recruitment and DHSS subsequently advised health authorities
to reconsider where they had reduced their intakes of
trainees. It is now clear that pressure on nurse
recruitment will continue for some time as the number of
eighteen year olds, the traditional age for entering nursing,
will fall by 12% during the 1980s, while the number of
employed nurses in the prime reproductive ages, and who may
therefore resign, is as high as 41%.

Nurse training has traditionally been on an
apprenticeship basis and learners are a major part of
the nurse workforce in hospitals. New supervisory and
regulating machinery for nursing is about to come into
force, and it remains to be seen what effects this will have.

This paper has identified some of the key issues in
medical and nursing manpower. A more detailed account would
have made reference to the increasing types and numbers
of technical and scientific staff that, as the technological
base of much of medicine expands, are seen as complementary
to the central disciplines of medicine and nursing. Suffice
it to say that in these cases there is normally a perceived
shortage.

Reference to 'perceived shortage' and 'common wisdom'
have been used in the preceding paragraphs as an indication
of the essentially subjective nature of manpower calculations.
Various norms and guidelines for different groups of the
NHS workforce have been promulgated nationally and locally.
Different suggested norms for the same group of workers
may coexist. They may be based on work study data, studies
of patient dependency or the informed but necessarily
partisan judgements of organisations and associations
representing a given profession. It is at this point necessary
to remember that despite some appearances, the NHS is not

a monolith. There is no direction of labour, as say
characterises the armed Services. It is rather, in some
respects, a federation of local health authorities:
each authority has considerable managerial autonomy
(outside basic pay rates which are set nationally) and must
make its decisions on manpower with due regard for local
circumstances including the local labour market.

One final observation related to the limited influence
local health authorities have on the determinants of
supply. This is not necessarily a major problem where any
training is given by the authority itself. But apart
from nursing, training is given by other bodies: the NHS
does not determine the capacity of training establishments
located in the further and higher education sectors and
subjected to the influence of the professions for which
they are training. The extreme case is the medical
profession.

Facilities

The key body in the development of new facilities is
the Regional Health Authority (RHA) which has as one of
its major functions responsibility for the Capital
Programme related to its strategic planning role. The
RHA will identify where within the Region new facilities
are required and thus draw up a Regional Capital Programme.
This has four elements: building schemes (new hospitals
and developments to existing hospitals); engineering schemes;
medical and scientific equipment; the fourth element is the
so called regionally managed services, the major one being
the Blood Transfusion Service, and in certain regions
containing a major conurbation the ambulance service for
that conurbation. The RHA also make available to each DHA
a 'block capital allocation', the purpose of which is
primarily to allow local management to carry out the
upgrading of facilities within the district. In turn the RHA
must submit the large projects (value over £5M) to DHSS
for approval. DHSS also control the starting date of such
schemes, but in other respects the management of the capital
programme is with the RHA. The DHSS, however, in keeping
with the nature of the central funding of the NHS,
determines how much each RHA will have to spend on capital
and indicates to the RHA its financial assumptions for planning
purposes.

Equipment and Supplies

Notwithstanding the 'federal' nature of the NHS referred
to above, there has been strong pressure from the early

days, not least from parliament, to exploit the potential
of the NHS for economies of scale through national and
regional contracts which local managers could buy off.
The nature of the central administrative arrangements have
varied over time, currently at central level there is a
special Supply Council, but the intent of centralising
contracting to the optimum level has been evident for many
years. The principal central functions are the placing of
central contracts for a range of items, the development
of specifications for items, and the identification and
assessment of 'best buys' particularly in the new
technologies.

     At Regional level a Regional Supplies Officer (RSO)
places contracts when no national contract has been made.
He works closely with the Regional Works Officer and his
staff in respect of the purchase of equipment associated
with capital schemes, and also with the Regional Scientific
Officer who is a key influence in the allocation to
districts of major new equipment and in giving professional
technical advice. At local level there is an Area Supplies
Officer who manages the supplies and stores function,including
buying off contracts fixed nationally or regionally, placing
local contracts, making individual purchases and maintaining
contact with local users of supplies.

     Pharmaceuticals are dealt with separately from other
supplies at national, regional and local levels but a
similar philosophy underpins the management of this
function. Constitutionally as a 'crown service' the NHS
does not have to observe the patent laws. This has generated
a wary relationship between the government and the pharmaceutical
industry. Voluntary price agreements and limits to the
amount of advertising to the medical profession have emerged
from this.
A distinction must be drawn between pharmaceuticals
purchased under contracts for hospitals and drugs dispensed
as an adjunct to general medical practice. In the latter
case the drugs prescribed by a GP are dispensed by a
private pharmacist or independent contractor
- who normally links his dispensing with a retail shop
selling a range of nonprescription drugs, cosmetics and
other goods. The pharmacist receives a handling charge
plus the cost of the drug prescribed. While drugs prescribed
in hospital are free to patients, in primary care the patient
makes a contributory payment for each item prescribed.
The GP has complete discretion - or clinical freedom - to
prescribe as he sees fit in the patient's interest but his
prescribing is monitored and, if well above the average in
his locality, he may be called to account. In hospitals

managerial efforts are made to contain costs although
this has proved difficult in practice. Hospital
pharmacists may be empowered to substitute generic
for brand named drugs when prescribed by junior medical
staff, but this rests on an agreed list being reached
with the medical staff. One other current controversy
concerns the balance locally between local production
and buying from the industry. Official policy currently
favours purchasing but hospital pharmacists may be
keen to develop their production activities.

Supply of Knowledge

     Inevitably the biggest source of knowledge
in the overall organisation is the countless contacts
between all health workers in managerial, service provision,
and support roles. The contracts fostered by personal
relationships, can be a specific intention by one party
to seek information or opinion from another, or can
occur as a by-product of a meeting. One notable characteristic
of the organisation is the frequency of meetings for
consultative, negotiating and other purposes that can
involve the personnel of one or more authorities.
Secondly, there is an ever burgeoning professional
literature both national and international which may be
directed to one occupational group or span several.
It is the outcome of such contacts which creates the climate
of knowledge and expectations. But within that climate
and in fulfilling its organisational role, DHSS assumes
a major responsibility. The principal bureaucratic method
employed is the issue of guidance circulars and policy
statements, which can vary in tone from consultative
(inviting comment) to exhortary to directive, but which
are all intended to influence the NHS whether in organisational
practices or policy thinking. DHSS will also, from time
to time, set up expert committees of personnel drawn from
the NHS, with particular terms of reference and subsequently
consult on their recommendations before deciding what
action to take. DHSS has considerable resources of its
own in terms of professional and technical staff who
maintain close contact with their peers in the NHS and there
is a well established practice of using advisory committees
and individual advisers and consultants.

     In recent years DHSS has also assumed a role in
commissioning research normally in areas which are thought
likely to have policy or other more immediate implications

for health services. There is a limited amount of intra-mural
research by DHSS research staff. The research function
is shared by other public funded bodies notably the
Medical Research Council and the Social Services Research
Council both of which sponsor particular research units
and fund individual projects. Recent developments suggest
a lessening role for DHSS as research sponsor and an
increasing role for MRC. The MRC and other research councils
have normally been seen as less concerned with research
of immediate relevance than a government department would
be, but such distinctions are difficult to sustain. In
any event the overall situation at the moment is that there
are likely to be less research funds available from public
sources for the foreseeable future, which adds new
importance to the lesser but still important role of the
voluntary foundations in sponsoring health services research.

Financial Resources

As indicated earlier, the NHS dominates the health
sector and is centrally funded essentially from general
taxation. Expenditure on private sector health services
has previously been estimated at perhaps 2% of the NHS
budget. Estimates will vary according to what items are
included and more recent estimates would allow for the
recent acceleration in private health insurance for hospital
and specialists services cover and a matching increase in
the bed capacity of the private sector. The number with
insurance is expected to rise to some 6 million (out of a
total UK population of 56 million) and the number of private
hospitals providing acute care could rise to over 200 but
these are nearly all small institutions and should be seen
against 147.000 acute hospital beds in the NHS.

The allocation to the NHS is negotiated each year
(together with a projection forward for a further 4 years)
by DHSS with the Treasury as part of an annual survey of
public expenditure and is formally approved by parliament.
Estimates presented to and voted on by parliament will be
in broad categories. For managerial purposes the important
breakdown by DHSS is between allocation to Regional Health
Authorities and Family Practitioner Services (FPS). The FPS
budget is regarded as "open ended" in that expenditure is
ultimately heavily determined by the actual claims made by
the independent contractors for their services or allowable
expenses. A global sum for these expenditures is set nationally;
it is at the local level that the budget is in practice open-
ended.

Family Practitioner Services

General Medical Services

- cost nearly £700 million
- about 22,000 GP's
- about 21,000 full-time equivalent ancillary staff
  (nurses, receptionists, etc.) employed by GP's
- over 180 million patient consultations per year
- 30 million home visits by the GP

General Pharmaceutical Services

- cost to public expenditure about £1,100 million
- charge income about £90 million
- exemptions for the elderly, the chronic sick, children,
  pregnant and nursing mothers, low incomes, war pensioners.

General Dental Services

- cost to public expenditure nearly £350 million
- charge income about £120 million
- exemptions for children, pregnant and nursing mothers, low-
  incomes
- about 12,500 general dental practitioners
- about 30 million courses of dental treatment

General Ophthalmic Services

- cost to public expenditure about £90 million
- charge income about £40 million
- exemptions for children, low incomes
- about 8,000 opticians and doctors
- about 8.5. million sight tests
- about 5½ million glasses etc. dispended

Hospital and Community Health Services

Money

- net current   £7,555
- capital   £ 630
- charge income (about 2/3 from private
  patient payments)   £   65

Manpower

Nearly 800,000 whole-time equivalent staff are employed.
Actual numbers of staff are larger because many are part-
timers. Major groups are:

```
- doctors and dentists                          38,000
- nurses                                       367,000
- professional and technical staff
  (therapists, lab technicians etc.)            65,500
- ancillaries                                  173,000
- ambulance staff                               18,000
- works and maintenance                         26,000
- administrative and clerical                  106,000
```

Physical Resources

- about 2,000 hospitals. A third of these were wholly or
  partly built before 1900: 85 major new projects are in
  the pipeline.
- 147,000 acute beds (over 40% occupied by people over 65)
- 139,000 beds for the mentally ill and handicapped
- 57,000 geriatric beds
- 19,000 obstetric beds

Breakdown of Expenditure

(a) by type of input (per cent)

-   74 goes on salaries and wages
-    3 goes on drugs
-    6 goes on medical and surgical equipment etc.
-   5½ goes on food, laundry, linen, furnishings, crockery,
     cleaning materials
-   3½ goes on fuel and water
-    4 goes on common services (e.g. rates, telephones)
-    2 goes on estate management, equipment, etc.
-    2 goes on vehicle and transport costs.

(b) by function (per cent)

-   10 goes on medical and dental services
-   31 goes on nursing services
-   12 goes on medical and surgical supplies and drugs
-    8 goes on medical support services e.g. investigative
     tests and therapy
-   19 goes on catering, laundry and domestic services
-    9 goes on medical records, administrative and clerical
     support, and miscellaneous services
-   11 goes on estate management (maintenance, boilers, etc.)

(c) by type of service (per cent)

-   55 on general and acute hospital services, including ambu-
     lances
-    6 on obstetric services

-   16 on services for the mentally ill and handicapped
-    3 on services mainly for children e.g. health visiting
-    2 on prevention and other community health services
-   12 on services specifically for the elderly
-    6 on administrative and support services

Activities

- nearly 4.1. million acute in-patient cases
- over 28 million acute out-patient attendances
- over 13 million accident and emergency attendances
- about 240,000 geriatric in-patient cases
- nearly a quarter of a million geriatric out-patient attendances
- 139,000 beds and 1.6 million out-patient attendances for the mentally ill and handicapped
- about three-quarters of a million obstetric in-patient cases
- 3.8 million obstetric out-patient attendances
- over 8,000 health visitors attending about 3.7 million cases a year
- nearly 14,000 district nurses attending about 3.2 million cases.

In contrast, allocation to RHAs and from RHAs are
fixed and cash limited, i.e. the proposed allocation
has an assumed inflation rate over the following
financial year. Health authorities cannot any more expect
any supplementary allocations above their given budget
and of inflation runs at a higher rate than assumed in the
calculation, each authority must trim accordingly. Recent
legislation has placed an obligation on health authorities
not to spend over their cash limits. Cash limits are a
source of anxiety to NHS treasurers and their management
colleagues. The prospect for the foreseeable future is that,
while it will continue to be a challenge to contain
expenditure within the limit, the prospect of a falling
national rate of inflation should ease the pressure to some
degree.

A major innovation in financial resource allocation to
Health Authorities in 1976 was the refinement and full
application of a systematic basis to the allocation according
to a concept of need. This was the application of a formula
put forward by an expert committee - Resource Allocation
Working Party (RAWP). The RAWP formula formally replaced
a method which evolved from 1948 whereby the original bid
from a Region and subsequently each year, last year's
allocation was increased to take into account inflation
and real growth. By the late 1960s this approach was seen
to produce dramatic differences in per capita expenditure
in the NHS between regions which could not be convincingly
explained away. Initial steps were taken to remedy these
differences in the early 1970s for hospital expenditure and
the development of a comprehensive formula was a logical step
after the 1974 Reorganisation. Essentially the RAWP formula
depends on population with account being taken of the
population structure and the use of Standard Mortality
Ratios (SMRs) as a proxy for morbidity. The SMR factor
and the absence of any other factor to take into account
socio-economic circumstances that could affect health
status, generated the most heated debate initially.
Subsequently, the RAWP approach appears to have been grudgingly
accepted even at the most local level of using the formula
to allocate from Regions to Districts.

Thus overall control of financial allocations does not
present a problem. The 6% of GNP spent on the NHS is
determined politically as part of government strategy
on public expenditure. RAWP and cash limits determine
allocations to health authorities. Where control is much
weaker is in the actual pattern of expenditure: here the key
is the concept of 'clinical freedom'. Successive governments
have indicated to the medical profession that they have

no intention of interfering with their professional practice.
Thus within a NHS system clinicians have more freedom than
in other seemingly more 'liberal professional' settings.
Clinicians interpret their clinical freedom as an obligation
to themselves and the individual patient rather than also
embracing a responsibility to the local NHS as an organisation
and to the local community. Thus there is virtually no
effective control of the countless individual clinical
decisions that in the end consume the resources made available.
Concepts such as medical audit, peer review and the like are
perhaps not quite so vehemently rejected as they once were,
and there is perhaps some sign of an increase of interest
in management and efficient use of resources among clinicians
but it is too soon to say that the climate has changed
fundamentally.

Lastly despite the dominance of central funding, some
reference should be made to voluntary effort. There is a
well established tradition of voluntary organisations
involved in giving service and raising funds as well as
acting as pressure groups nationally and locally. The balance
between these three forms of activity varies from organisation
to organisation: those who are significantly involved in
giving service are often in receipt of grants from DHSS
or health and local (social service) authorities. As a
proportion of the total NHS budget these grants are negligible
but they are seen as ensuring a valuable complement or
supplement to the statutory services.

One other voluntary activity deserves mention - local
funding raising initiatives: these have been increasingly
common in recent years particularly raising funds for advanced
medical equipment (such as CT scanners) which have not been
given a high priority for purchase with public funds but
which have been given extensive media exposure suggesting
that they are major medical breakthroughs. It is extremely
difficult for a NHS authority to decline gifts of this kind
although the running costs and eventual replacement costs
will quickly exceed the capital gift. The effect of acceptance
is necessarily to reorder the priorities of that authority.

Patterns of Health Services

Reference has already been made to the structural
split in the UK medical profession between the GPs and
the Specialists - a distinction sometimes characterised as
'the GPs have the patients and the Consultants have the beds'.
This split was well established before the NHS but the initial

organisational structure of the service contributed to
its perpetuation. In the context of the NHS the aphorism
can be explained as follows: each patient registers with a
GP of his choice and the GP is free to accept or reject a
patient who seeks to register with him. For as long as the
patient is registered with a GP he is entitled to assistance
and advice from him. The agreement can be terminated by
either party, but while it runs (and relatively few people
change their GPs unless they move to a different locality)
the GP is paid a capitation fee each quarter year whether
or not he has attended the patient.

The model GP-Patient relationship emphasises its continuing
nature, the GPs knowledge of the patient's family, social and
economic circumstances, and the GPs potential to act as
counsellor and educator and practitioner of preventive
as well as curative medicine. The GP's contractual
commitment is in effect limited to responding in a reasonable
and professionally competent way to requests for assistance
in the patient's home or in the doctor's premises - 'general
medical services' in the language of the legislation. Each
GP interprets his role beyond fulfilling his contractual and
legal obligations in his own way, but there is some evidence
of a shift among many GPs towards the assumption of a wider
role. Partly this has been achieved by financial inducements
from government, partly as a result of an intellectual and
professional resurgence through the efforts of the Royal College
of General Practitioners which has emerged as a significant
professional association, and the establishment of
Departments of General Practice in University medical schools
with dynamic leadership.

The self-image of the GP and his interpretation of his
role is important because it is the GP who decides whether
to treat the patient with his own resources or to refer the
patient to a hospital based specialist, either for a consultative
outpatient appointment or in emergency directly as an inpatient,
or if necessary he can request a domicillary visit by the
specialist to the patient's home. The GP is thus a switching
mechanism who can bring all the resources of the NHS into
operation on behalf of his patients. He has no immediate
financial incentive to refer his patient or not to refer.
His inclination and his professional judgment make him the
gatekeeper to the more expensive resources of the hospital.
Before the resurgence of general practice he was seen
increasingly as a postbox - an intermediary a patient had to
deal with before gaining access to 'real' medical care
available only at a hospital since consultant specialists
will only see patients referred to them by a GP (or another
specialist).

Part of this resurgence of General Practice is manifested in the emergence of the Primary Health Care Team which has evolved since the late 1960s. There is no standard uniform composition of a PHC team but it normally involves a GP with associated nursing personnel, perhaps a social worker and secretarial and receptionist support. The existence of a team as a small organisational unit, and increasing direct access to diagnostic facilities at hospital without first referring to a specialist have provided a major boost to primary care. However, it is still exceptional for GPs to have direct access to hospital beds and Consultants have a strong sense of ownership to the beds allocated to them to manage as a unit or 'firm' of their speciality. While the GP can act as an advocate for his patient, the split between GP and Consultant does set limits to the potential of primary care.

Thus the patient's normal pathway into the NHS is straightforward. Indeed his only other general first point of contact would be the Accident and Emergency Department of a major general hospital. Although these are not intended to provide a primary care service, in practice they would be reluctant to turn patients away even when it would be equally appropriate to go to a GP.

Continuity and Integration of Care

Increasingly, over the last two decades or so, there has been a frequently expressed concern over the absence of continuity of care and integrated care. The separation of general medical practice from the specialties is both a strength in that the patient has a known point of first contact with the system and a weakness in that once the GP has referred a patient there is no necessary motive or incentive for him to take any interest in the patient until the Consultant discharges him or the patient seeks care for a quite separate condition. Closer contact between GPs and Specialists is clearly a necessary pre-condition for continuity of care. It is probably true that over time there has been an increase in mutual respect and awareness of each others role assisted by such innovations as local postgraduate medical centres which have provided a potential common meeting ground. Nevertheless, the sense of separate identity is strong and it would be over-optimistic to expect dramatic change in the short run.

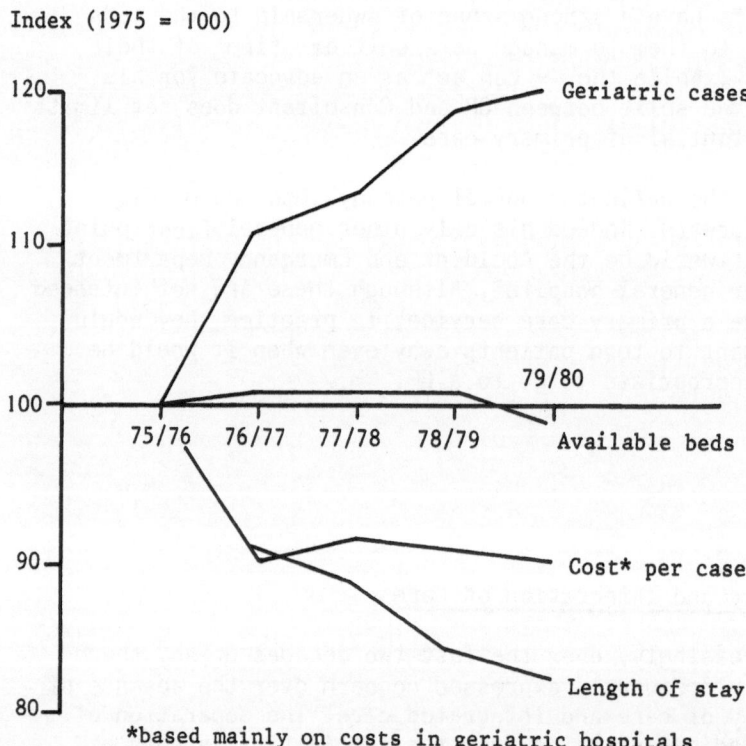

Figure 4   In-Patient Services for the Elderly 1975-80.
           *Based mainly on costs in geriatric hospitals.

Concern over the lack of integrated care has been most intensely felt over the high dependency groups such as the frail elderly, the mentally ill and the handicapped. Individuals in these care groups have a variety of health, social and economic needs which are the responsibility of different professions and agencies. The 1948 NHS structure was seen to exacerbate the problems of people with complex needs and this was one of the reasons identified for seeking a reorganisation which would bring all services within one structure. While such a change could not guarantee integrated care, it was argued that at least it would further the possibility. The 1974 Reorganisation placed emphasis on top management's responsibility for promoting integrated care through such means as comprehensive planning, effective co-ordination within the organisation and fully developed liaison machinery at planning, operational management and case levels with other agencies. One specific innovation which was officially encouraged was the setting up of planning teams for these care groups. The membership of these teams was to include all disciplines with an involvement whether they were employed by the NHS or another agency; representation of the consumer and of relevant voluntary organisations was also encouraged. The rationality of such an approach was believed to be self-evident, but despite individual successes reported, the overall impression after some years of experience was that again the sense of separate identity and territorial rights was too strong to assume that a smooth form of joint working could be easily achieved. One particular problem facing top management was how far teams could be left to develop their own ideas and how far they needed close supervision. In the one case management risked losing control; in the other, the team could lose its motivation.

It is perhaps tempting to be too dismissive of these efforts and, while progress may be slow and indeed patchy, the introduction of formal machinery and recommended approaches to joint working will have had some influence on those working with the high dependency groups mentioned earlier. There are well established statutory services which can be identified as the responsible parties and thus be invited to collaborate. Although other groups such as the alcoholics and drug addicts and (in a different category) ethnic minorities from different cultures, have equally complex needs, there is a time lag before a public social responsibility for such groups becomes widely accepted and a policy response and organisational arrangements developed. In these circumstances, any care for these groups is more likely to be provided by voluntary organisations and hence there will be considerable variations in the quality and scale of what is available.

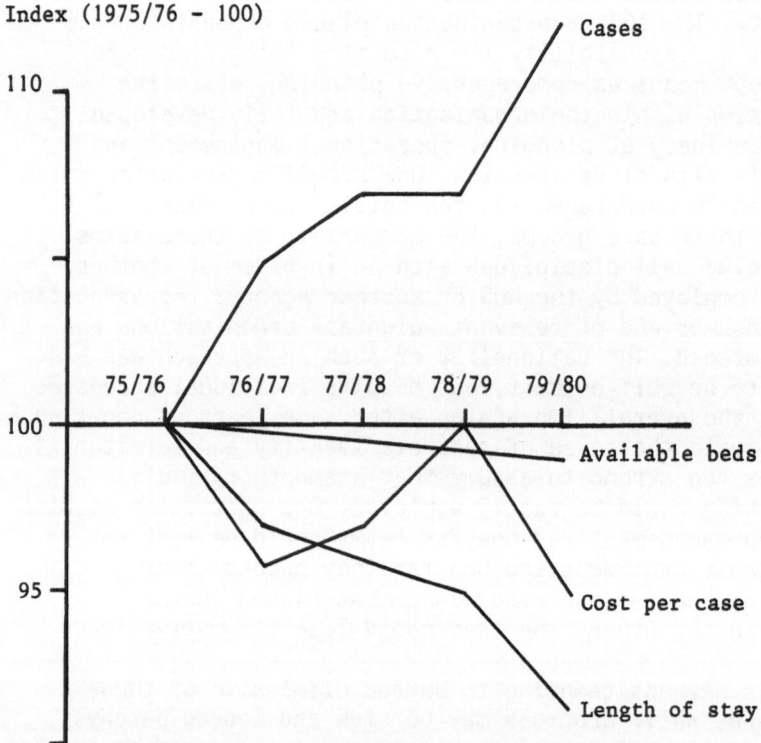

Figure 5   Acute Services 1975-80 (All in-patients in acute, mainly
           acute, and partly acute hospitals).

Health Promotion

Despite the comprehensive character of the formal objectives of the NHS, in terms of the actual deployment of resources and the issues which are most likely to generate public and professional interest, the emphasis remains on the acute curative services despite the declared intentions of successive Ministers over decades to shift the balance to long stay care or even prevention. If anything, prevention and health promotion have had less political, public and professional attention than long stay care. Occasional Ministers have identified with particular prevention issues such as smoking, but this has never been sustained.

The counselling educative role of the PHC team is almost always subordinated to their curative role. Health Education Officers have been employed by health authorities since 1974 and there is a national Health Education Council but neither at local or national level has it been possible for this emergent profession to engage the support of more powerful forces. The current pressure on financial resources with NHS has made prevention and health promotion attractive to politicians as low cost alternatives to curative services but the inevitability of disappointed expectations which will follow in due course could then make it even more difficult for health education and health promotion to command attention when resources are being allocated.

Indeed, in the end these activities may suffer from a lack of proven efficacy and thus compare very unfavourably with the services related to environmental control both generally and in the workplace which would seem to be carried out for the most part efficiently and without external attention.

Scottish Variations

What has been described is largely common to England and Scotland. There are, however, some Scottish variations, largely relating to the politico-administrative structure, which should be identified.

Scotland is a country within a country. It is part of the United Kingdom of Great Britain and Northern Ireland governed in the same manner as other parts of the Union. It differs, however, from other parts of the U.K. in certain significant respects. For example it has its own legal system. The health, education and other local government systems are

separate, being administered centrally by a group of
functional Departments known as the Scottish Office, this
being the Ministry of the Secretary of State for Scotland.

Responsibility to Parliament for the Scottish Health
Service rests ultimately with the Secretary of State for
Scotland (a senior Cabinet Minister). To run the health
services the Secretary of State has a department or ministry
to help him - the Scottish Home and Health Department.
Responsibility is further delegated for the more local
organisation and planning of health services in the
15 areas into which the country is divided to Health Boards.
In the largest health board areas there is a further division
into health districts, the officers of which are responsible
to the health board for the provision of health services within
the district boundaries.

At each major level of management - central, area,
district - there are appropriate medical and other health
professional advisory committees to advise management. There
is also advice from Health Councils representing lay and
consumer interest.

Additionally there is a Scottish Health Services Planning
Council which is the principal source of planning and policy
advice to the Secretary of State and which contains
representatives from all 15 health boards, SHHD and the four
universities with medical schools. In association with
SHHD it is involved in the evolution of national policies,
priorities and plans. Within the broad framework of these
national policies and plans the 15 health boards carry out
more detailed planning and policy formulation for their
respective areas. There is no planning function at district
level, district officers being charged solely with the
efficient organisation and running of the various services.

At both national and area level programme planning
groups do the planning for the different client or patient
care groups.

Additionally in Scotland there is what is known as the
Common Services Agency. The CSA has no policy formulation
function whatsoever but is intended to provide a variety
of services both to SHHD and the 15 Health Boards. It is made
up of a number of quite heterogeneous divisions. There are
divisions responsible for operational services such as
Ambulances and Blood Transfusion which are organised on an
all Scotland basis. There are divisions in support of
administrative and planning functions such as Supplies,

Information Services,Building and Legal Services,again
acting on an all Scottish basis,as well as Health Education
and Communicable Disease Control where the remit is also for
the whole of Scotland.

The basic rationale behind the Common Services Agency
is that it is a mechanism for providing those services
which are most efficiently organised on a national rather
than an area basis but which nevertheless do not need to be
within the framework of a central government ministry.
The operational policies of the divisions of the CSA are
determined by a Management Committee made up of
representatives of SHHD and the 15 health boards areas.

Another difference which can be noted in Scotland,
but is difficult to define in formal structural terms,
is a greater sense of over-all coherence in the system.
This is certainly due partly to the smaller scale of operations
involved for a population of only 5.2 million but probably also
reflects different cultural factors. There is a distinct
tendency for there to be much more informal communication
and decision-making between the different bodies and levels
before the formal processes are invoked. This often makes
reaching consensus easier.

The diagram below summarises this system.

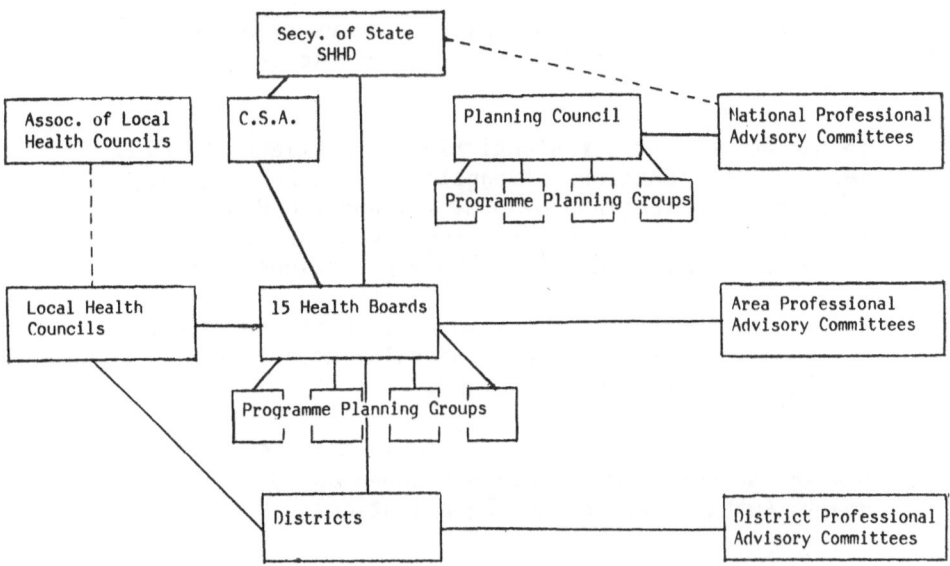

Figure 6   Health Planning Structure of Scotland:

In relation to what has been said in the earlier sections of this paper, the Scottish variations from what pertains in England might be summarised thus:

(I)   Scotland has two additional central bodies not found in England; the Scottish Health Services Planning Council and the Common Services Agency. Both of these appear to perform useful and effective functions in the Scottish context.

(II)   The N.H.S. administrative structure has been less complex and therefore has given rise to fewer problems than in England. The administration also seems to be facilitated by easier and more informal communication between the different parts of the system.

(III)   Scotland appears to have developed a more complex professional advisory system than in England and though this has had advantages a disadvantage has been to produce additional inertia in the decision making processes.

(IV)   In medical manpower Scotland does not have the same problems as England in that traditionally it has over-produced doctors, the surplus being exported to England or overseas. There are also liaison mechanisms which work well between the Deans of the four medical schools and senior officers of SHHD and the Health Boards to ensure symbiotic arrangements between the Universities and the N.H.S.

(V)   In the absence of a regional tier in Scotland S.H.H.D. performs the functions of capital allocation normally carried out in England by D.H.S.S. and the Regions with the technical aspects being taken care of by the Building Division of the Common Services Agency.

(VI)   Private medicine is insignificant in Scotland with both the public and the medical profession much less attracted to it than in England.

(VII)   There is a resource allocation formula in Scotland similar to the RAWP formula and known as SHARE (Scottish Health Authorities Revenue Equalisation).

REFLECTIONS: the lessons of organisational change and corporate planning.

A concern to improve the management of the NHS has been evident from the 1960s when there were a number of steps taken
to expand the provision of management education to various
groups of staff. The 1960s also saw a political commitment
to replace the old physical facilities by a network of all
purpose district general hospitals. But the debate about the
role of the NHS, its priorities and objectives became locked
into a debate about organisation structure and the most
appropriate form of managerial hierarchy. Awareness of a
range of shortcomings in the NHS generated a search for the
organizational panacea which became a matter of "salvation
by structural solutions".
From about 1966 to 1973 when legislation was passed by parliament, the whole intellectual and political debate focussed
on alternative blueprints on NHS structure. The bargaining
between interest groups to preserve their positions after
change overshadowed the underlying rationale of creating an
environment which increased the probability of the planned
development of sensitive services relevant to people's perceived needs. In this debate, the NHS was to some extent prey
to fashionable thinking concerning large complex organisations
in both the public and private sectors of the economy and the
virtues of comprehensive structures and corporate management
and planning systems, with perhaps an understatement of the
conditions to be satisfied, the mix of incentives, and sanctions to secure the full implementation of agreed objectives.

The 1974 NHS structure was certainly less than the
original architects of change would have wished, but its
features were for the most part defensible in terms of
the objectives behind them. It was perhaps unfortunate that
there was a delay of two years before the complementary
innovation of a corporate planning system was introduced.
It was even more unfortunate that the implementation of the
system was accompanied by an economic crisis that made the
stability of financial assumptions an impossibility, and hence
struck at the credibility of planning. It was an error of
administrative judgement that did not link together from the
outset the planning system and the resource allocation formula
innovation (RAWP). It was a grave error of psychological and
tactical judgement to increase the detailed documentation
required by the Centre which appeared to run counter to the
original decentralisation philosophy and which generated
either suspicion, resentment or indifference at local level.
But that is too pessimistic a picture. The introduction of the
planning system at least established that:

- the identification of objectives should precede questions
  of resource requirements and organisation changes required;
- it is essential to involve all interested parties whose

co-operative participation is necessary to the full
implementation of a policy;
- Ministers felt obliged to publish explicit statements
  of their policies to be interpreted and adapted
  by NHS authorities;
- articulated plans should identify where responsibility
  for action lay - at national, regional or local levels.

Unfortunately, the errors may have had greater impact
than the successes but at least corporate planning in the NHS
has survived a change in political climate and a change in
government. Indeed it is to be hoped that in future planning
and the understanding of the NHS as a complex system will
be governed by a new realism. They will in short be governed
by modest expectations, by the realisation of the importance
of bargaining between interest groups rather than an imposed
rationality,which cannot be sustained by a simple form of
planning, which limits both the issues covered to keep
it within the span of human comprehension and the documentation required to prevent a bureaucratic paperchase

In summary, the history of the NHS system has many
accounts of error and misjudgement, but its future development still warrants modest optimism

Select Bibliography

The above narrative may be supplemented by reference to the
following publications.

Barnard, K. and Lee K. (eds.) (1977) Conflicts in the NHS
(Croom Helm)
Barnard K. et al, NHS Planning: An Assessment, in Hospital and
Health Services Review, August and September 1980
Barnard K. et al, Patients First: Intentions and Consequences
(1980), Nuffield Centre for Health Services Studies, University
of Leeds
Brown R.G.S. et al, New Wine Old Bottles, (IHS, Hull)
Chaplin N. (e.d.) (1982), Health Care in the United Kingdom
(Kluwer Medical)
DHSS, (1972), Management Arrangements for the Reorganized NHS
DHSS, (1976), NHS Planning Manual
DHSS, (1979), Patients First (1980), Health Circular HC
(80) 8
DHSS, (1981) Health Notice HN (81) 30, Health Building Proce-
dures
DHSS, (1982) Health Circular HC (82) 6, The NHS Planning System
Harrison S.R. Consensus Decision Making in the National Health
Service a review, Journal of Management Studies (forthcoming)

Haywood, S. and Alaszewski. A. (1980) <u>Crises in the NHS</u>
(Croom Helm)
Jackson P.M. (e.d), (1981), <u>Government Policy Initiatives</u>
1979/80 (RIPA, London), (Chapter 9, Health Services)
Klein, R. (1981) The Strategy behind the Jenkins non-
strategy, <u>B.M.J.</u>, 28, 3, 1983.
Laing, W. (1982) A Mixed Economy in Health: Public/Private
Collaboration? in RIPA Report, Summer 1982 (RIPA, London)
Levitt, R. (3rd Edition), <u>The Reorganized NHS</u> (Croom Helm)
Long A. and Mercer G. (1981), <u>Manpower Planning in the NHS</u>
(Gower)
McLachlan G. (ed.), <u>The Planning of Health Services: Studies
in 8 European Countries</u> (WHO, Copenhagen), (Chapter 1 - Scot
land)
Razzell E.J. (1980) Planning in Whitehall, <u>Journal of Long
Range Planning</u>, February 1980
Royal Commission on the NHS (1979) <u>Report</u>, Cmnd. 7615 (HMSO)
Willcocks, A.J. (1967), <u>The Creation of the NHS</u> (Routledge and
Kegan Paul)
WHO (1981), <u>Health Services in Europe</u> (3rd Edition), vol. 2
(WHO, Copenhagen), Country reviews and statistics.

THE CHANGING HEALTH SERVICES SYSTEM IN GREECE

Anna Ritsataki

Centre of Planning & Economic Research

Athens, Greece

Introduction

The fact that the new socialist government in Greece
has only been in power for ten months, and during this time
has initiated basic changes in practically all sectors of the
economy, means that we are very much in a state of flux.
Although certain general policy directions are readily apparent,
at this very preliminary stage, there is obviously still considerable uncertainty as to the manner in which these changes
will be implemented. For this reason, and due to the limited
time available, the following is simply an attempt to pick out
some of the points suggested in the outline for country papers.

Values, Goals, Objectives

In recent years, there has been a considerable convergence
of the views of the main political parties concerning the development of the health services. The previous conservative government,
whilst not advocating the immediate establishment of a national
health service, or the restriction of the private sector, stated
in its five-year development plan, that it would not take any
measures which would hinder the future establishment of a national
health service, and immediately prior to its fall from power, had
prepared a draft law providing for the setting up of a central and
regional planning system, the regionalization of the hospital
services, the establishment of a network of health centres as out-
posts of the hospitals, and the development of a family doctor
system.
The present government aims at giving priority to the health
sector, and considers "a high level of health, as of education, as

a basic precondition for the achievement of other aims of socio-
economic development.

In the preamble of the Law for the establishment of a Central
Health Council, which was recently passed by parliament, the
government stated its objectives for the health sector as being:
"The socialisation of health, which entails free medical, pharmaceutical, hospital and preventive care for all Greeks, the abolition
of private clinics and all special advantages in the provision of
medical and hospital services .... The problem in the health field
is mainly organizational, and the operational structure has been
defined, for the provision of a high level of health for all the
population, regardless of their social and economic position, and
their place of residence."

"None of the parameters which secure the public health, will be
for profit. It will be ensured that:
a) Every citizen will be freed from the insecurity and anxiety
   caused by sickness, handicap and old age.
b) Every citizen will have the same right to equal and high level
   treatment and social care.
c) The implementation of the above will be the sole responsibility
   of the state."

The achievement of these objectives is obviously going to
present some problems, given the very strong private sector and the
commercial attitude of many physicians, the very advantageous health
benefits enjoyed by certain groups of the population such as bank
employees and civil servants, the lack of public confidence in the
existing public health services, and the present, extremely uneven
geographical distribution of health facilities and personnel.

Political, Administrative and Geographical Configuration

Greece has a population of almost 10 million, nearly 45%
of which live in the regions of Greater Athens and Salonika.
The country is characterized by its rugged mountains and small
islands, and the provision of health services to population scattered in small communities, in remote mountainous regions and
small islands, presents considerable problems.

At the present time, the country is divided into 53 nomoi
or counties, each of which is administered by a "nomarch" appointed
by the government. The "nomarch" is the direct representative of
the central government at the county level, and is head of a service
composed of representatives of the various Ministries, that is,
a kind of miniature replica of the central government at the local
level. This centralized system is very similar to the old prefect
system in France.

There is no second-level authority between the central government and the community and municipal authorities. Until now, the community and municipal councils had very little responsibility, mainly due to their lack of financial resources.

The present government has already started to organize County Committees in which representatives of local professionals, cultural political and other organizations, have been given the opportunity to discuss local problems and make proposals for future development. Within the periode of the 1983-87 development plan, second-level local authorities are to be established at the level of the present counties or possibly a lower level, and at a later date, regional authorities are also to be created. One of the main aims of the socialist government is to decentralize power, and devolve responsibility to the local authorities.

Structure of the Health Services Systems

At the present time, the private health sector in Greece is very strong. Of the approximately 60.000 hospital beds in the country, about 46% are in state hospitals, 14% in voluntary hospitals, and 40% in private clinics, and the distribution of total hospital discharges between these hospitals is very similar.

In recent years, the rate of increase in beds in private clinics has been faster than in state hospitals, mainly due to the development of private maternity clinics, and a number of "luxuoury" clinics in the Athens area. The socialist government has stated that it will not allow the opening of new private clinics, and has offered to buy out private clinics and their equipment. A number of private clinics have apparently already moved to take up this offer.

The development of private clinics has not been planned, licenses simply being awarded to clinics which met certain standards. Although many of these private clinics are extremely small (30% of private units have 1-20 beds, 28% 21-40 beds), and are poorly staffed and equipped, they have been able to operate through contracts with the state insurance organisations. The government's intentions with regard to the use of private clinics by insurance organizations, have not yet been announced. Since they account for such a large proportion of total beds, however, obviously a number of the private clinics will have to continue in operation for some time.

The government intends to regionalize the hospital services, and to coordinate them with primary care. The state hospitals in each county are to be run as one administrative unit. A system of

Health Centres is to be gradually established throughout the
country, both in rural and urban areas.
The Health Centres are to be extensions of the hospital services, and will work in close cooperation with a system
of family doctors.
The Health Centres are also to be responsible for preventive
health care, which is presently not on an organized basis.
In addition, they will have a welfare section, which should
facilitate to some extent, the coordination of the health and
welfare services, which is presently totally lacking.

The extent of physicians' private practice is unknown. Until
the present time, doctors employed in state hospitals and by the
public insurance organizations, worked an average of 3-5 hours a
day in those positions and held other positions in private clinics
or ran their own private surgeries. There is no estimation as yet
of this multiple position holding and, therefore, of the extent
of the private sector.
The lack of confidence in the medical profession has facilitated
the growth of the private sector since patients tend to consult
two or even more physicians even for relatively simple health
problems. In addition, it is widely held that many physicians refer
patients unnecessarily to colleagues, on a reciprocal basis.

Until now coordination between primary and hospital care has
been negligible, and there was no provision for follow-up of discharged hospital patients. Admission to hospital, particularly the
hospital of a patient's choice, was frequently gained by the back
door through private consultation with a doctor employed in the
hospital, or even political string-pulling.
The planned health system is expected to regulate hospital admissions, and to create the necessary links between the different types
of health care.

At the present time, a draft law is being prepared which will
deal with the employment of doctors in state hospitals. Doctors are
to be employed on a full-time and exclusive basis in state hospitals, that is, doctors who join the state hospital system, will
be prohibited from engaging in private practice. This is the
second law to be brought to Parliament by new government, concerning the health sector. It is difficult to assess, however, whether
this is an indication of the priority which is to be given to hospital care, or whether it is more the result of the acute problems which have been created in the state hospitals, by the premature dismissal of all state hospital physicians last December,
and their subsequent re-engagement on a temporary basis, until the
new, full-time system comes into effect.

Health Resources Production

Education and Supply of Manpower

The supply of physicians in Greece in relation to the population is one of the highest in the world. There are about 420 inhabitants per physician at the present time for the population as a whole, though this ranges from about 250 inhabitants per physician in the county of Attica, to over 2.500 in Kastoria.

With regards to the total number of physicians produced, there has been no real attempt at coordinating education and manpower planning, partly because of the pressure for admittance to the medical schools. As for all university education in Greece, a fixed number of students are admitted to the first year of the medical schools each year. To avoid this obstruction, however, many young people attended foreign universities, particularly Italian universities, returning to Greece after one or two years' study abroad, to gain admittance to the Greek universtiy in the second or third year, with obvious detrimental effects on the quality of education, due to the enormous pressure on the facilities. Sporadic attempts have been made to prevent this, and to restrict the numbers admitted to Greek medical schools. So far, popular pressure against such restrictions has been so strong, that even the military dictatorship was unable to enforce them.

Th new educational policy which has been announced, provides that whilst the number of undergraduates admitted to the university will be fixed, and entrance will continue to be on the basis of competitive examination, the individual student will be given an unlimited number of chances to try to gain entrance to higher education.

There has been no planning for the specialisation of physicians, which has resulted in very great shortages in certain specialities, and an abundance of physicians in what were considered to be the more lucrative specialities (gynaecologists and pediatricians in particular). Although for a number of years it has been stressed that family doctors should be responsible for primary care and should play a key role in the delivery of health services, there has been no corresponding change in medical education. It is mainly the pathologists who undertake the role of the G.P. or family doctor in Greece, though their training for this is not really adequate.
Furthermore, even though the aging make such a disproportionate use of the health services, geriatrics is not a recognized speciality in Greece.

In the future, admission to medical schools for undergraduate and post-graduate education, is to be geared to the
needs of the health services, and special Medical Education
Councils are to be set up, which will be given this responsibility.

There is a very great shortage of trained nursing personnel.
Of hospital nurses, for example, only 36% have three or four
years' training, 18% have a one-year nursing certificate and 46%
are practical nurses with a low level of general education. In
general hospitals there are 2.9 beds per nurse (including practical
nurses) and in neuropsychiatric hospitals the corresponding
figure is 5.8. Since the trained nurses are employed mainly on the
morning shifts, in small hospital units in particular, there may
be no trained nurses on duty in the evening.

The lack of nursing personnel is mainly due to the low social
standing of the nursing profession, and the failure to take strong
measures to rectify this. Until recently, the nursing schools were
compelled to repeat their entrance examinations two and three
times a year, in an attempt to fill availabe places. More recently, improvements in the level of nurses pay, the rising level of
unemployment amongst young people, and the very strong competition for university placed, have attracted more, and better
qualified candidates to the nursing schools.

The pre-election plans of the present government provide
for the setting up of a nursing school in every hospital. A
number of other measures would probably have to be taken to fill
such schools, notably provision for the part-time employment
of nursing personnel, which is important for married women. So
far, however, the nursing profession does not seem to have made
its voice heard very strongly in the health planning process,
which is presently dominated by physicians.

The Technical Colleges (KATE) provide training for medium
level hospital administrators, medical visitors and laboratory
assistants. There was no coordination, however, between the
output of the KATE and positions in the health services, with
the result that graduates from these colleges have found themselves without employment. These colleges are, however, to be
abolished, and to be replaced by a new type of technical
training. Details of the proposed new technical schools are not
yet available. It is to be presumed that they will take into
account the very great need for trained middle level personnel
in the health services.

Training for hospital management is not available in Greece.
For a wide variety of other vital personnel also, such as, maintenance technicians for medical equipment, dieticians, physio-

therapists and ergotherapists, speech therapists, medical
social workers and medical secretaries, there is no
specialised training, or this is available only on a very
limited scale.

Facilities, Equipment, Supplies

The existing serious regional disparities in the provision
of health facilities, indicate that they have not been planned
according to need.

There has been no rational planning for equipment. As in
many countries, hospital equipment has been seen as a status
symbol, and with this incentive, medical directors with sufficient influence, have managed to obtain expensive equipment,
some of which is simply sitting in hospital basements. On the
other hand, certain relatively simple instruments which are
standard equipment in many other countries, are not to be seen
in Greek hospitals, since doctors have not been trained in their
use.

In fact, there is no reliable estimate of hospital equipment.
In 1978, Control Data tried to make a rough evaluation, but of
2000 questionnaires which were sent to hospitals throughout the
country, only 24 were completed, and even for those, there was
considerable difficulty with the categorisation of equipment.
At the present time, the University of Patras is trying to make
a rough estimate of hospital technology.

The use of drugs is considered to be excessive. Until the
present time, drug manufacturing was entirely in the hands of
private enterprise, and the market is dominated by the multi-
national corporations. The government has recently announced the
setting up of a state drug industry, and is preparing a national
Drug List which will limit the number of drugs eligible for payment by the public insurance organisations. It is hoped that these
and other measures, will curb the excessive use of drugs.

Supply of "Knowledge"

Health research is carried out mainly in the universities and
university hospitals, and has been largely oriented towards clinical research. The first attempt at planning research was made by
the previous government, which set up a special service for
planning and directing the public financing of research, but this
was not considered to be successful. The present government has
announced that it will finance research which might contribute to
achieving the aims of the five-year development programme, and has
invited the submission of such research proposals. This seems to
have initiated an interest in some of the medical schools in

research into the need for, and possible patterns of provision
of health services.

Financing the Health Services

Only 3.5% of GNP is spent on the health services in Greece,
though this should be increased by an unestimated amount paid
by individuals to physicians, sometimes in the form of a "tip"
or "little envelope" as it is known in Greek. This type of under-
the-counter payment can be considerable, but does not appear as
doctors' income, and therefore, also escapes taxation.

The rural population and urban workers are covered for primary and hospital care by their separate insurance organizations,
and together they account for almost 70% of the population. There
are about another 50 insurance organizations covering various professional groups, and hospital care in particular is covered
mainly by payments from insurance organizations. Since hospital
payments do not always cover the costs, the public hospitals are
subsidised regularly by the state.

Cost control in the state hospitals is simply a question of
balancing the books, rather than questioning the necessity of the
expenditures. In some cases, this type of control seems to be
more expensive than allowing for a certain amount of "loss" of
hospital supplies.

There are no evaluation committees in the hospitals, to
assess the necessity of medical actions, or the length of hospitalization, although hospitalization is generally considered
to be prolonged for other than medical reasons (lack of nursing
homes, domiciliary health services, family reasons etc).
Neither have the medical unions and associations assumed a role
of self-supervision and control, as in certain other countries.

The previous government, prior to its fall from power, was
preparing a new system of payment for hospital care, with the
aim of shifting more of the cost to the insurance organizations,
and reducing the level of state subsidies to public hospitals.
A certain amount of cost control in private clinics is exercised by state regulation of the price which may be charged for
in-patient care, both to insurance organizations and individuals.
There have also been sporadic attempts at curbing expenditure on
drugs.

Physicians' fees, when they are contracted to, or when these
are reimbursed by insurance organizations, are also regulated.
Fees charged by physicians in their private surgeries, are, however, unregulated, and a large proportion of physicians' income
from private fees, is considered to remain hidden.

Patterns of Health Services Delivery

The broad outline of the pattern of health services delivery which is planned, has been referred to above. Details of its implementation are not yet available.

In the first few months of the new government, one change which has been noticeable, is the lack of attention so far given to preventive care. This could of course, simply be a temporary reaction to the previous government's policy. The former government waged a comparatively successful campaign against smoking, which was remarkable since Greece is a tobacco producing country, and also carried out intensive campaigns against over-weight, and to encourage blood donations. It is to be hoped that the present situation is simply a temporary lapse, and that the fight against smoking in particular, will be quickly renewed.

One preventive measure which is perforce being implemented, is the fight against air-pollution in Athens. This has reached such dangerous levels, that private cars are to be banned from the centre of the city during certain hours.

Health Services System Management

Unfortunately, there is a complete lack in Greece, of personnel, trained to run a health services organization, and this is probably going to be one of the main obstacles to the rapid and effective implementation of the proposed changes.

The Ministry of Health and Welfare does not have sufficient qualified personnel for health services planning and management. The main force behind the new plan for health, is a party-political group of young doctors, who are active in the medical associations.

Up to the present time, the medical associations have been mainly concerned with the professional interests of their members, which have frequently been pursued to the detriment of the level of care offered. For example, the terms of service which have been successfully negotiated by certain groups of physicians with regards to hours of work, have been detrimental to the accessibility and continuity of primary care, and have led to a marked reduction in the proportion of physicians' home visits. The medical associations now seem to be setting themselves up as the watchdogs of both the interests of the people and of their own members. The possible conflicts in such a position are obvious.

With regard to planning, as was mentioned above, a law has just been passed, which sets up a Central Health Council. This Council is to be composed of 24 members, who are representatives of the medical associations, workers' and farmers' unions,

medical schools, the local authorities, civil servants, the
nursing profession, the urban and rural insurance organizations,
the drug industry and the Ministry of Health and Welfare. The
Council is to be appointed for a three year term, and is to
have responsibility "for the planning and definition of the
general aims and policy directions, the formulation of national
policy in the health sector, and the submission of proposals to
the Minister of Health and Welfare". To facilitate the work
of the Council, a Secretariat is to be set up in the Ministry of
Health and Welfare, to be staffed by personnel of the Ministry,
and other special scientific personnel. Similar councils are to
be set up later, at the county, or second-tier, local authority
level.

References

1. "Health" Report of Working Group, 1976-80 Development Plan,
   Centre of Planning and Economic Research, Athens, 1976.
2. "Regionalization of Health" Report of Working Group, Regional
   Development Plan, Centre of Planning and Economic Research,
   Athens, 1980
3. Gana, A. Tsiatas, E. (rapporteurs) "Study for the Improvement
   and Organization of the Primary Health Care of the Urban
   Population", Greek Productivity Centre (ELKEPA), Athens, 1982.
4. Gana, A. in collaboration with Zervou, F. "Drugs - Level of
   Prices and Price Policy", Centre of Planning and Economic
   Research, Athens, 1982
5. Kaczka, E., Ritsatakis, A. "An Analysis of Administrative
   Functions, and The Cost-Structure of Selected State Hospitals
   in Greece", Centre of Planning and Economic Research, Athens,
   1974
6. "Policy Directions of the Panhellenic Socialist Movement",
   PASOK, Athens, 1977
7. "Contract with the People", Panhellenic Socialist Movement,
   Athens, 1981

PARTICIPANTS

1. Mr. D. Affeld
   Bundesministerium für Arbeit und Sozialordnung
   Postfach 14 02 80
   D-5300 Bonn
   Federal Republic of Germany

2. Prof. A. Bariletti
   Universita di Roma
   Facolta di Economia e Commercio
   Instituto di Scienza delle Finanze
   Via del Castro Laurenziano, 9
   00161 Roma
   Italy

3. Mr. K. Barnard
   The Nuffield Centre for Health Services Studies
   71-75 Clarendon Road
   Leeds LS 29 PL
   United Kingdom

4. Prof. J.E. Blanpain
   Centre for Hospital Sciences
   Catholic University of Leuven
   Vital de Costerstraat 102
   B-300 Leuven
   Belgium

5. Prof. P.B. Checkland
   Department of System Sciences
   University of Lancaster
   Lancaster
   United Kingdom

6. Prof. L. Delesie
   Centre for Hospital Sciences
   Catholic University Leuven
   Vital de Costerstraat 102
   B-3000 Leuven
   Belgium

7. Mrs. Prof. A. Crichton
   Department of Health Care and Epidemiology
   Faculty of Medicine
   The University of British Columbia
   Mather Building
   5804 Fairview Building
   Vancouver B.C. V6T 1W5
   Canada

8. Dr. A. Haidekker
   C.H.S. Müller Unternehmensbereich
   der Philips GmbH
   Goldröschenweg 43
   P.O.B. 650347
   2000 Hamburg 65   *
   Federal Republic of Germany

9. Dr. A.S. Häro
   Department of Planning and Evaluation
   The National Board of Health
   Siltasaarenkatu 18 A
   00530 Helsinki 53
   Finland

10. Mr. G.B. Hirsch   Co-Chairman
    Health Management Consultants
    7 Highgate Road
    Wayland, Massachusetts 01778
    U.S.A.

11. Dr. B.M. Kleczkowski
    Resource Group
    Division of Strengthening of Health Services
    World Health Organization
    1211 Geneva 27
    Switzerland

12. Mrs. B.J. Kostrewski
    The Centre of Information Science
    The City University
    Northampton Square
    London EC1V 0HB
    United Kingdom

13. Dr. J.F. Lacronique
    Ministère de la Santé Publique
    20-Rue d'Estrées
    75700 Paris
    France

14. Prof. H.J.J. Leenen
    Institute for Social Medicine
    University of Amsterdam
    Heerengracht 520
    Amsterdam
    The Netherlands

15. Prof. A.C.J. de Leeuw
    Interfaculty of Management Sciences
    State University Groningen
    Pleiadenlaan 10
    9704 CB Groningen
    The Netherlands

16. Mr. R. Lopes dos Reis
    Bureau of Studies and Planning
    Ministry of Social Affairs
    Av. Alvares Cabral, 25
    1200 Lisbon
    Portugal

17. Dr. I.S. Luculescu
    Country Health Programming
    WHO Regional Office for Europe
    8, Scherfigsvej
    DK-2100 Copenhagen
    Denmark

18. Mr. F. Marziale
    Public Health Division
    Council of Europe
    BP 431 R6-67006 <u>Strasbourg</u> Cedex
    France*

19. Prof. D.M. Pendreigh
    Department of Health Administration
    Usher University
    Warrenden Park Road
    <u>Edinborough</u>
    Scotland
    United Kingdom

20. Mr. J.P. Poullier
    Social Affairs Division
    Organization for Economic   Cooperation & Development (OECD)
    2 Rue André Pascal
    75775 <u>Paris</u> Cedex
    France

21. Prof. M. Rahmi Dirican
    Department of Public Health
    Medical School
    Bursa University
    <u>Bursa</u>
    Turkey

22. Mrs. A. Ritsataki
    Hellenic Republic
    Centre of Planning and Economic Research
    22, Hippokratous Street
    <u>Athens</u> 144
    Greece

23. Prof. M.I. Roemer
    Department of Health Administration
    University of California
    <u>Los Angeles</u>
    California 90024
    U.S.A.

24. Mr. H.T. Skaug
    Ministry of Social Affairs
    P.O. Box 8011
    N. <u>Oslo</u> 1
    Norway

25. Mrs. Prof. A.R. Somers

    Department of Community Medicine
    College of Medicine and Dentistry of New Jersey
    Rutgers Medical School
    University Heights
    Piscataway, New Jersey 08854
    U.S.A.

26. Prof. A. van der Werff   Chairman
    Staff Bureau for Policy Development
    Ministry of Health and Environmental Protection
    Dr. Reijersstraat 8-12
    2260 AK Leidschendam
    The Netherlands

    Department of Policy Sciences
    Faculty of Social Health Sciences
    State University Limburg, Maastricht
    Postbox 616
    6200 MD Maastricht
    The Netherlands

27. Dr. A. Weber
    Health Information
    WHO Regional Office for Europe
    8, Scherfigsvej
    DK-2100 Copenhagen
    Denmark

28. Dr. G. Wennström
    Planning Department
    National Board of Health and Welfare
    10630 Stockholm
    Sweden

29. Dr. W. Wils
    Wils Systems Analysis
    Pr. Mariannelaan 246
    2275 BN Voorburg
    The Netherlands

---

*

    Unable to attend.

II. Rapporteurs

    Mrs. H. Emanuel
    Staff Bureau for Policy Development
    Ministry of Health and Environmental Protection
    Dr. Reijersstraat 8-12
    2260 AK Leidschendam
    The Netherlands

    Dr. J.W. Hartgerink
    Staff Bureau for Policy Development
    Ministry of Health and Environmental Protection
    Dr. Reijersstraat 8-12
    2260 AK Leidschendam
    The Netherlands

    Mr. A.W.M. Meijer
    Department of Policy Sciences
    Faculty of Social Health Sciences
    State University Limburg, Maastricht
    Postbus 616
    6200 MD Maastricht
    The Netherlands

    Mrs. Dr. I. Mur
    Department of Policy Sciences
    Faculty of Social Health Sciences
    State University Limburg, Maastricht
    Postbox 616
    6200 MD Maastricht
    The Netherlands

    Mr. H.J. Roelants
    Centre for Health Information
    Postbox 14066
    3508 SC Utrecht
    The Netherlands

III. Editors of the Conference Proceedings

    Mr. K. Barnard
    The Nuffield Centre for health Services Studies
    71-75 Clarendon Road
    Leeds LS 29 PL
    United Kingdom

Dr. Ch. O. Pannenborg
Staff Bureau for Policy Development
Ministry of Health and Environmental Protection
Dr. Reijersstraat 8-12
2260 AK Leidschendam
The Netherlands

IV. Secretary of the Conference

Dr. Ch. O. Pannenborg

V.   Secretariat of the Conference

Mr. A.H. Zwennes
Mrs.M.A. Lamerée
Mrs.M.A. Holtmans   Staff Bureau for Policy Develop-
Mrs.F.W.M.A. Jans de Haan   ment

Mr. P. Käuderer   Department of International
                             Affairs

                             Ministry of Health and Environ-
                             mental Protection
                             Dr. Reijersstraat 8-12
                             2260 AK Leidschendam
                             The Netherlands

Adequacy of services, 122, 132
  see also assessment
Aims in planning, 72-73
  see also specific names of
      countries, objectives
Analysis
  general approach, 47-60
  summary, 15-16
  see also systems analysis
      (approach)
Area Health Authorities, UK, 317,
    318
Assessment of services
  adequacy, 122, 132
  cost-effectiveness, 124-125,
    132
  effectiveness, 123-124, 132
  efficacy, 122-123, 132
  efficiency, 125, 132
  overall system, 130-133
  sub-systems, 130-133
  see also Economics: Financing
      health service systems

Bed capacity, Turkey, 260
Belgium health service systems,
    18, 249-254
  delivery patterns, 253
  development, 10
  economics, 252-253
  historical facts, 249-250
  management, 253-254
  patient referral, 20
  primary care, 18
  resources production, 251-252
  structure, 22, 250-251

Blue Cross and Shield plans, USA,
    235-236, 242-243
Budgetting *see* Financing health
    services

Canada, health service system,
    283-309
  budget global - Quebec, 104
  coherency, 302
  Constitution Act sequelae, 305
  coordination of sources,
      299-300
  cost control, 297-299
  decision-making processes,
      303-304
  delivery patterns, 300-302
  economics, 296-300
  elderly and handicapped, 301
  entrepreneurs, 285-286, 291
  Federal/Provincial government
      factors, 288-290,
      291-292, 294, 297-298,
      300, 305
  geographical features, 283
  government bureaucracies,
      286-287
  Health Charter for Canadians,
      287-288
  historical features, 10, 284
  interest groups, 285
  knowledge and technology, 296
  legislation and regulation
      revision, 304
  management, 302-303
  objectives, 284-288
  political configuration,
      288-290

Canada, health service system
       (continued)
  primary health care, 302
  promotion of health, 301-302
  research, 154
  resources production, 294-296
  structure, 22, 290-294
    coordination with health
        sector, 291-293
    with other related sectors,
        293-294
Care
  assessment, 120-133
    see also Assessment:
        Cost-effectiveness
  classification of policies, 57
  continuity and integration, UK,
        335-339
  delivery patterns, 8-9, 55-56
    see also Delivery patterns
  integration, 28
  planning for, 93-95
  policy and finance, 107
  preventive/curative, 26-27
  primary see Primary care
  research into, 153-163
    see also Research and
        development
  summary, 26-29
Centralization, Netherlands, 194
Classification of systems see
        Types
Clinical autonomy, 20, 22
Cohesion, stimulation
        Netherlands, 191-195
Common Services Agency, Scotland,
        339, 342
Communication see Information
        systems
Communicative planning, 84
Community Health Councils, 318,
        322
Components of health service
        systems, 50-59
Consumer participation, 30-31
Cost
  benefit analysis, 116-117
  control mechanisms, Norway, 278
    Canada, 297-299

Cost (continued)
  effectiveness, systems, 115-135
    assessments of adequacy, 122
      cost-effectiveness, 124-125
      effectiveness, 123-124
      efficacy, 122-123
      efficiency, 125
      overall system, 130-133
    definitions, 116-119
    distinction from cost benefit
        analysis, 116
    evaluation, 119-121
      systems, 125-130
    and scientific communities,
        119
    sequences of sub-systems, 120
    terminology, 116-119
    vs. effectiveness, 118

Decentralisation, Netherlands,
        195-198
    see also Regionalization
Decision making
  Canada, 303-304
  Italy, 217-218
  Turkey, 269
Definition, health services
        system, 50-51
Delivery patterns
  Belgium, 253
  Canada, 300-302
  Greece, 355
  Italy, innovations, 216-217
  Norway, 279-280
  UK, 333-335
  USA, 239-241
Demand, determinants, summary,
        22-23
Denmark, health service
        development, 9
    see also Scandinavian countries
Determinants of health, 47-49
Diagnostic-related groups
        costing, 103-104
Disjointed incremental type of
        planning, 81
District Health Authorities, UK,
        319-321
Drugs, 53
  Belgium, 252
  UK, 326-327

Economics, health services,
        54-55, 56, 57
    Belgium, 252
    Canada, 296-301
    France, 172-177
    input/output/budgetting/
        financing summary, 35-38
    NATO countries, summary, 20-21
    Norway, 278-279
    Turkey, 264-265
    USA, 234-239
    Western World, 8
Education
    Belgium, 251-252
    Canada, 294-295
    of consumers, 33
    Netherlands health service,
        198-199
    Norway, 277
    of personnel, 96, 97
    USA, 230
Effectiveness of services,
        123-124, 132
Efficacy of services, 122-123,
        132
Efficiency of services, 125, 132
Employers' payments, USA, 236-237
Equipment, 33
    Belgium, 252
    Canada, 295-296
    France, 171-172
    Greece, 353
    Turkey, 264
    USA, 230-233
Evaluation, 40, 86-87, 119-121
Evolution, health service
        systems, 49-50

Facilities, 33, 51
    Canada, 295
    Greece, 353
    Turkey, 264
    UK, 325
    USA, 230-233
Family Practitioner Committee,
        UK, 317, 322
Family Practitioner Services, UK,
        329
Financing health service systems,
        99-114
    Belgium, 252-253

Financing health service systems,
        (continued)
    budget, global,
        France, 105
        Netherlands, 105
        Quebec, 104
    Canada, 296-301
    care policy, 107
    cost stage, 108
    and decreasing hospital costs,
        107-108
    definition of concepts, 100-101
    diagnostic related groups
        costing, USA, 103-104
    French health care system,
        173-174
    hospital budgets, Sweden,
        105-106
    Italy, 211, 212, 213-216
    maxicaps, USA, 104
    mechanisms, 108-109
        appraisal of differences,
            109-113
        classified, 102
    Netherlands, 186, 188-189, 190
        restructuring, 198
    new tariff proposals, France,
        103
    Norway, 278-279
    parallel situations, 107
    patient profile stage, 109,
        110, 112
    price per inpatient day, 101,
        103
    production stage, 108
    prospective rate ('batch'
        rating), 103
    reimbursement of costs, 101
    reimbursement stage, 108, 110
    resource allocation working
        party, UK, 105-106
    supply or demand, 109
    Sweden, 198
    Turkey, 264-265, 269
    UK, 198, 328-333
    USA, 222, 223, 224-228
        hospital costs, 231-233
    Western World, 8, 54
        mechanisms, 35-38
    see also Cost-effectiveness

Finland health service
        development
  information services, 141-151
  structure, 21
  team approach, 27-28
  see also Scandinavian countries
France, health service system,
        167-177
  budget global, 105
  consumption, 174-176
  economics, 172-177
  finances, 173-174
  future, 177
  general objectives, 167
  health system structure,
        169-170
  political orientations, 176-177
  political structure of country,
        168
  private patients, 21
  production of services, 170
  research, 154
  systems view, summary, 18
Fund raising for specific
        equipment, 17
Germany, health service
  development, 10
  research, 154, 155
    systems approaches, 156,
        159-163
Greece, health service system,
        347-356
  Central Health Council
        establishment, 348
  delivery patterns, 355
  development, 10
  education, 351-353
  equipment, 353
  facilities, 353
  geographical distribution,
        348-349
  hospital care, 350
  knowledge, supply, 353-354
  management, 352-353, 355-356
  manpower, 351-353
  nursing training, 352
  objectives, 347-348
  patient referral, 20
  political configuration, 348
  primary care, 350

Greece, health service system
        (continued)
  private sector, 349-350
  public/private sector, 21
  resources production, 351-354
  structure, 21, 349-350
  supplies, 353

Health insurance system, cost
        effectiveness, 128-129
  see also insurance schemes
Health Services Act, Netherlands,
        191-193
Health service systems
  and reorientation, summary,
        15-46, 61-68
  Western World, 7-14
    developments, 9-10
    environment, 7-8
    economic and financial
        factors, 8
    major concerns, 7-9
    management, 9
    methods, 12-13
    objectives, 11-12
    patterns of delivery, 8-9
    production, 8
    structure, 8
    WHO concern, 10-11
Health Systems Agencies (HSAs,
        USA), 231-232
Holland see Netherlands
Hospital care
  France, 18
  Greece, 350
  Holland, 18
Hospital Facilities Act,
        Netherlands, 191-193

INAM medical care benefits,
        Italy, 205-207
Indicative planning, 84-85
Information systems, 39-40,
        137-151
  communication, 149-150
  content, 143-144
  cooperation, problems, 142-143
  coordination, problems,
        138-139, 142-143
  criteria, 140-141

Information systems (continued)
  data processing and analysis,
     148
  definitions, 139
  locations, 142
  management, 138-139
  methodological aspects, 146-148
  modelling, 144-146
  National Health systems,
     139-140
  organization, 142
  presentation, 149-150
  reliability, 148-149
  services, 140
  strategic problems, 141
  structure, 142
  systems analysis, 144-146
  technical aspects, 146-148
  training, 150
Input control, 35, 36, 131, 132
Insurance schemes 128-129
  Belgium, 251
  Italy, 17, 205-207
  USA, 227, 236, 242-243
Integration of care, summary,
     28-29
Italy, health service system,
     203-218
  decision making, 217-218
  delivery patterns, innovation,
     216-217
  development, 10
  expenditures, 213-216
    control, measures, 215-216
    growth, 214-215
  formed structure, 21, 209-211
    P.S.N. planning framework,
       210
    S.S.N., 209-210
      financial/administrative
         circuit, 211, 212
  health care, 211
  health insurance funds, 17,
     205-207
  management problems, 217
  manpower policy problems, 213
  objectives, 203-204
  political system, 204-205
  pre-reform system, 205-208
    general structure, 205-206

Italy, health service system
     (continued)
  pre-reform system (continued)
    insurances schemes, 17,
       205-207
    operational characteristics,
       206-208
    private/public sectors, 208
    social services, 211
    staff organization, 217

Joint Consultative Committees,
     UK, 317-318

Knowledge and technology, 34-35
  Canada, 296
  France, 172
  Greece, 353-354
  Turkey, 264
  UK 327-328
  USA 233-234

Long range planning, 93-94

Management
  Belgium, 253-254
  Canada, 302-303
  of information services,
       139-140, 143-144
  Italy, 217
  Norway, 280-281
  UK, 352-353, 355-356
  USA, 242-243
  Western World, 9
    summary, 38-41
  see also Evaluation: Infor-
       mation systems: Monitor-
       ing: Planning: Research
Manpower, 51
  Belgium, 251-252
  Canada, 294-295
  France, 170-171
  planning, 91-98
    care, 93
    changing personnel
       requirements, 91-93
    comprehensive policy, 95-96
    to improve manpower
       situation, 93-95
    long range, 93-94, 97
    optimum utilization, 96-97

Manpower (continued)
  summary, 32-33
  Turkey, 262-264
  UK, 321-325, 329-331
  USA, 230
Maxicaps, costing, 104
Medicaid, 227, 229, 242
Medical profession system in cost
        effectiveness, 127
Medicare, 227, 229, 237, 242
Medigap, 227
Mixed-scanning type of planning,
        82-83
Monitoring, 40

NATO countries, health services
  Civilian Science Programme, 5
    see also Systems Science
        Panel
  common problems, summary, 16-19
Netherlands, health service
        system, 179-202
  budgetting experiments, 105
  centralization, 194
  coherence, 186
    stimulation, 195-200
  cohesion, 191
    stimulation, 195-200
  counter pressures, 192-195
  dangers, 192-195
  decentralization, 195-198
  decision-making, 199
  development, 9-10
  economic control, summary, 20
  education, 198-199
  finances, 186, 188-189, 190
    restructuring, 198
  insurance systems, 186-187
  interdependency, 183-184
  payment, systems, 187
  physicians, autonomy, 20
  private/public funding, 186-187
  research, 154, 200
  routine control, 191
  systems theory, 180-183
  unbalanced growth, 189
North America see Canada: United
        States of America
Norway, health services, 271-281
  delivery patterns, 279-280

Norway, health services
        (continued)
  development, 9
  economics, 278
  finances, 278-279
  general features, 271
  historical features, 272-274
  management, 280-281
  national health plan, 272
  objectives, 272-275
  political configuration, 275
  primary health care, 280
    running by municipalities,
        272
  regionalization proposed, aims,
        274
  resources production, 277-278
  structure, 21, 275-277
    coordination with health
        sector, 275-277
    intersectoral coordination,
        277
  see also Scandinavian countries
Nursing training and manpower
  Greece, 352
  UK, 324

Objectives
  of conference, 11-13
  in planning, 73-74
  see also specific countries
Outcome indicators, 131, 132
Output control, 35, 36-38, 131,
        132

Pay beds, 21
  see also Private sector
Personnel requirements, changing,
        in health services, 91-93
Physicians, 51
  autonomy, 20, 22
  distribution, Turkey, 260
  patient referral to, 20
  social training for, 32
Planning, 39, 69-89
  aims and objects, 72-74
  complexity, 78
  decisions and control, 75-76
  dimensions, 71-72
  evaluation, 86-87

Planning (continued)
  human growth and progress, 75
  levels, 73-74
  management, 91-98
    see also Management
  need for, 69-70
  problems of theory, 74-76
  rationality, 78-79
  role, 70
  substantive versus procedural,
      74-75
  types, 79-86
    see also Types
  uncertainty, 78
Portugal, health service
      development, 10
Preventive health care, 27-28
  Norway, 279
Primary care
  availability, 18
  Belgium, 250
  Canada, 302
  Greece, 350
  Norway, 272, 280
  UK, 18, 333-335
  USA, 240-241
Private
  insurance schemes, 128-129
    Belgium 251
    Italy, 17, 205-207, 208
    USA, 227, 235-236, 242-243
  patients, 21
  sector, Greece, 349-350
    Netherlands, 186
    and public sector
        relationships, 21
    UK, 311-312
    USA, 224-226, 227, 234,
        235-236, 237
Problems
  in planning theory, 74-76
  solving process, 61, 63-64
    see also Decision-making
Programmes, organization, 53
Promotion of health, Canada,
      301-302
  UK, 339
Public sector, 17
  Italy, 208
  and private sectors
      relationships, 21

Public sector (continued)
  USA, 224-226, 234-235

Rational-comprehensive/synoptic
      type of planning, 80-81
Regional Health Authorities, UK,
      313, 316-317, 325
Regionalization, 21, 31-32
  Canada, 290-294
  Netherlands, 192, 196-198
  Norway, 274
  UK, 313, 316-317
Reorientation of health services
  initiatives, summary, 26
  summary, 15-46
  systems approach, 42-44, 61-68
    methods, 45-46, 61-66
    tools, 67-68
Requirements for effective
      systems, summary, 24-25
Research and development, 40-41,
      153-163
  France, 172
  into convergency and divergency
      of variables, different
      countries, 162-163
  into general system dynamics of
      health care, 160
  Netherlands Health Service, 200
  systems approaches, 155-156
    marginal importance, 156-159
    strengthening, 159-163
    see also Systems approaches
  UK, 154
  USA, 154, 155
  Western countries, trends,
      154-155
Resources,
  coordination and distribution,
      NATO, summary, 18-19
  development, 51-53
  production, 32-35
    see also Education: Equipment:
      Facilities: Hospital
      care: Manpower:
      Physicians: Requirements:
      Supplies: Training: and
      specific countries

Scandinavian countries, health
        care system, summary, 17,
        18
    see also specific countries
Scotland, health services,
        339-342
    structure, 341-342
Scottish Health Services Planning
        Council, 340
Sickness funds
    co-ordinating, 20
    Italy, 205-207
    Netherlands, 186-187
    see also Insurance schemes
Soft system methodology, 62-64
    see also Systems approach
Specialists, patient referral to,
        20
State control and inspection
        system, cost
        effectiveness, 129-130
Structure
    Belgium, 250-251
    Canada, 290-294
    consumer participation, 30-31
    France, 169-170
    Greece, 349-350
    information services, 142
    Norway, 275-277
    private sector, 29
    public sector, 29
    regionalization, 30, 31
    summaries, 21-22
    Turkey, 259-262
    UK, 311-313, 314-321
    USA, 223-227
Students attending paramedical
        schools, Turkey, 263-264
Supplies, 33
    Canada, 295-296
    Greece, 353
    Turkey, 264
    USA, 230-233
Sweden, health service
    cost, 91
    development, 9
    finances, 198
    hospital budgets, 104-105
    structure, 21
    see also Scandinavia

System-rational type of planning,
        83-84
Systems analysis (approach)
    and cost-effectiveness, 115-135
        see also Cost effectiveness
    and information, 144-146
    Netherlands health services
        system, 180-183
        decomposable system,
            181-182
        part systems, 180
    to reorientation, 42-44, 61-68
        methods, 45-46
        1950-1969, 62
        1970-1980, 62-64
        paradigms, 64-66
        planning, 76
        soft system methodology,
            1970-1980, 62-64
        tools, 67-68
    research, 155-156
        ambiguity of concepts,
            157-158
        deficits of knowledge, 157
        inadequacy for peculiar
            delivery patterns,
            158-159
        mix-up of programmatic and
            factual system
            contingency, 158
        strengthening system aspects,
            159-163
    Science Panel, chairman's
        address, 5
    USA, 160-163

Team approach, 27-28
Technology, 34-35
    France, 172
The Netherlands *see* Netherlands
Training of personnel, 96
    France, 170-171
Treatment, cost-effectiveness,
        124-125
    see also Cost-effectiveness
Trends, 59-60
    research and development,
        154-155
    USA, structure, 223-227

Turkey, health service system,
    255–270
  bed capacity, 260
  care providers, 259–261
  climate, 255
  coordination within health
      sector, 261
    within other sectors, 262
  delivery patterns, 265–268
  development, 10
  distribution of physicians, 260
  economics, 264–265
  equipment, 264
  facilities, 264
  geography, 255
  health situation, 256–257
  institutions, 265–267
  knowledge and technology, 264
  management, 268–269
  manpower, 262–264
  multipurpose health units,
      257–258
  national income, 256
  nationalization, 257
  objectives, 257–258
  policy decisions, 269
  population, 255–256
  problems, 258
  public care, 17
  resources production, 262–265
  structure, 259–262
  supplies, 264
Types
  of health service systems,
      56–59
  planning, 79–86
    communicative, 84
    disjointed incremental, 81
    indicative, 84–85
    mixed-scanning, 82–83
    rational-comprehensive/
        synoptic, 80–81
    styles, 79
    system-rational, 83–84

Unbalanced growth, Netherland
    Health Service, 189
Union influence USA, 237
United Kingdom, health service
    system, 309–346

United Kingdom, health service
    system (continued)
  Area Health Authorities, 317,
    318
  Community Health Councils, 318,
    322
  continuity and integration of
    care, 335–339
  delivery patterns, 333–335
  Department of Health and Social
    Security and NHS, 313–314
  development, 9
  District Health Authorities,
    319–321
  District Management Team, 318
  elderly care, 336–337
  equipment, 325–327
  facilities, 325
  Family Practitioner Committee,
    317–322
  family practitioner services,
    329
  finances, 198
  financial resources, 328–333
  general practitioners, 323,
    333–335
  health promotion, 339
  hospital-based specialties,
    323–324
  hospital and community health
    services, 329
  Joint Consultative Committee,
    317–318
  knowledge, supply, 327–328
  manpower, 321–325
  National Health Service,
    313–314
    administration, 314–321
    restructuring 1982, 319–321
  nursing training and manpower,
    324
  pharmaceuticals, 326
  political background, 309–311
  primary care available, 18
  private patients, 21
  private sector, 311
  regional health authorities,
    313, 316–317, 325
  regionalization, 21
  research, 154

United Kingdom, health service
      system (continued)
   resource allocation working
         party, UK, 105-106
   resources, 321-333
   Royal Commission on NHS, 1976,
         310
   Scottish variations, 339-344
   social security, 313
   statistics of activities, 331
   structure, 311-321
      reflections on, 342-344
   supplies, 325-327
   systems approach, 159-163
   system, summary, 17
   team approach, 27-28
United States of America, health
         services, 219-248
   coordination, 227-229
   coordination of all services,
         239-240
   costs, 230-233
   costs, rise, 223
   delivery patterns, 239-241
   Development, 10
   economic support, 234-239
   economics, 20
   education, 230
   employers' payments, 236-237
   equipment, 230-233
   facilities, 230-233
   financing, 103-104
   financing review, 222, 223,
         224-228
   general background, 220-221
   high technology, 239
   HMO's, 241
   inflation, 227
   institutionalization,
         increasing, 223
   international activity, 238-239

United States of America, health
         services (continued)
   knowledge and technology,
         coordination, 233-234
   low technology, 240
   management, 242-243
   manpower, 230
   objectives, 220-221
   patient referral, 20
   political configuration, 221
   political leadership, 242
   primary care, 240
      government role, 241
   private health insurance,
         235-236
   public programmes, 235
   publicly supported planning
         agencies, HSA's etc.,
         231-232
   research, 154, 155
   resources production, 230-233
   structure, 22, 223-227
   supplies, 230-233
   system, summary, 18
   systems approach, 160-163
   technology and knowledge,
         coordination, 233-234
   'third party' payment, 225
      associated with institutional
         care, 225
   union influence, 237
   voluntary health agency, 238

Voluntary health agency, USA, 238

Western World, health service
         systems, 7-14
   see also Health service systems
         and specific countries
World Health Organization (WHO),
         approach to health
         service systems, 10-11